Biology

SUBJECT DICTIONARIES

CHEMISTRY MINI DICTIONARY

★ A to Z for handy reference
★ Extensive cross-referencing
★ Explanatory illustrations
★ Entries keyed to National Curriculum Attainment Targets
★ Traditional and current chemical nomenclature

ISBN 0 245-60325-5
£2.75

PHYSICS MINI DICTIONARY

★ A to Z for handy reference
★ Extensive cross-referencing
★ Explanatory illustrations
★ Entries keyed to National Curriculum Attainment Targets

ISBN 0 245-60327-1
£2.75

HARRAP'S

Biology

MINI DICTIONARY

Edited by
John O.E. Clark
and
William Hemsley

HARRAP
London

First published in Great Britain 1992
by HARRAP BOOKS Ltd
Chelsea House, 26 Market Square, Bromley
Kent BR1 1NA

ISBN 0 245 – 60374 – 3

Typeset by Action Typesetting, Gloucester
Printed in Great Britain by
Richard Clay Ltd,
Bungay, Suffolk

Preface

Harrap's Biology Mini Dictionary has been devised for two main groups of readers. The first group consists of people whose daily work brings them into contact with biological terms. They may be practising biologists or – because of the nature of their jobs – non-biologists who nevertheless have to understand biological terminology. Readers in the second main category are students, and for them we have tried to meet several needs. Schoolchildren and students who are learning biology will find the Dictionary invaluable for checking the meanings of words that are a required part of the vocabulary of the subject. Students in allied disciplines, such as biochemistry, chemistry and medicine, can use it as a handy reference source for words from biology that are commonly employed in *their* subjects, but are so often taken for granted.

The overlap of biology with other disciplines posed a problem for the compilers of this Dictionary. There is a natural progression from biology to biochemistry and molecular biology, and then from these subdisciplines to chemistry. Similarly biology is a precursor of medicine. So where does biology stop? Certainly the main terms of biochemistry have to be included, as does some of the terminology of medicine – a biochemist and a medical student both have to know what a hormone is. The dilemma is shared by the committee that awards the annual Nobel Prizes, which have no category called 'Biology'. Some renowned biologists find themselves recipients of the Chemistry Prize, although most are awarded a prize in the category Physiology or Medicine (*see* Appendix III).

J.O.E.C. and W.T.J.H. – London, 1991

GCSE and National Curriculum

In order to help students using this book as a study aid, articles of a particular relevance at GCSE level are indicated by ■ at the beginning of the entry. In addition, many articles are coded at the end to key them into current National Curriculum Attainment Targets – for example [13/4/d]. The first number of the code gives the attainment target, the second the level of attainment and the final letter indicates the statement within that level.

Please note that this coding refers to the attainment targets for the Science Curriculum which are in force at the time of publication. A table at the back of the book (*see* Appendix IV) relates these targets to the *New* Attainments Targets (NATs), which are due to be introduced on 1 August 1992 in all key stages.

The NATs are designed to simplify the structure of the National Curriculum. There are five NATs, as follows:

NAT 1	Scientific Investigation
NAT 2	Life and Living Processes
NAT 3	Earth and Environment
NAT 4	Materials and Their Behaviour
NAT 5	Energy and Its Effects

A

■ **abdomen** *1.* Rear body section of an **arthropod**. *2.* Part of the main body cavity in a **vertebrate** that contains organs other than the heart and lungs. In mammals it is separated from the **thorax** by the **diaphragm**.[3.4.a] [3.5.e] [3.8.a]

aberration In biology, the change in the number or the structure of **chromosomes** in a cell. *See also* **mutation**. [4.6.a]

abiogenesis Theory of the origin of life which states that living organisms formed from non-living material, either the old-fashioned belief held until 150 years ago (*see* **spontaneous generation**) or the modern one that life-forms gradually evolved from simple cell-like collections of biochemicals.

■ **abortion** Spontaneous or induced termination of pregnancy, resulting in the expulsion of the foetus. A natural abortion is often termed a miscarriage. [3.4.c]

■ **ABO blood groups** Most commonly used classification system for human blood. The blood groups are named after the **antigen** types A and B in red blood cells. The groups are A, B, AB and O. Each blood type has corresponding **antibodies** in the blood **serum**. Group O contains neither antigen. Group A has type A antigens (and type B antibodies); group B has type B antigens (and type A antibodies); group AB has both antigens (and neither antibody); group O has neither antigen (and both antibodies). If blood from a donor is given to a patient, the patient's own blood must not have antibodies to the antigens in the blood that is given or agglutination rejection will occur. Group O is therefore a universal donor. [3.4.b]

abscisic acid Growth-inhibiting plant **hormone** that causes leaves to be shed (**abscission**).

■ **abscission** The shedding of a leaf, fruit or other part of a plant, caused by the formation of a layer of cork cells (abscission layer) at the base of the part which cuts off its supply of nutrients and water. It is brought about by abscisic acid.

■ **absorption** Taking up of matter or energy by other matter; *e.g.* digested food is absorbed into the bloodstream from the intestines.

■ **abyssal** Relating to the ocean floor (*e.g.* abyssal animals, or fauna).

accelerator In biology, a substance that increases the effectiveness of an **enzyme**.

■ **acclimatization** Way in which an organism adapts to a different or changing environment (*e.g.* the thin air at higher altitudes).

■ **accommodation** Ability of the **eye** to alter its focus by changing the curvature of the lens or, in the eyes of a few animals (*e.g.* cephalopods), changing the position of the lens relative to the retina. [15.6.c]

■ **acellular** Not made up of **cells**; in particular, describing an organism that is not divided into separate cells. *See also* **coenocyte**.

■ **acetylcholine** (ACH) White hygroscopic crystalline organic compound. It is important in the nervous system of many animals, where it transmits impulses between **synapses** of nerves; a **neurotransmitter**. After transmission it is broken down by the enzyme cholinesterase.

acetylsalicylic acid Alternative name for the analgesic drug **aspirin**. [3.4.b]

ACH Abbreviation of **acetylcholine**.

■ **achene** Small dry **indehiscent** fruit that contains a single seed, as in many nuts, dandelion 'seeds' and the 'pips' on strawberries.

■ **acid** Member of a class of chemical compounds whose aqueous solutions contain hydrogen ions (H^+). Solutions of acids have a **pH** of less than 7 (pH = 7 is neutral; a pH of greater than 7 is alkaline). Strong acids dissociate completely (into ions) in solution; weak acids only partly dissociate. An acid neutralizes a **base** to form a **salt**.

acidosis Condition in which the **pH** of the blood falls below its normal value of 7.35. It can have various causes, such as a breathing difficulty which prevents the lungs from removing sufficient carbon dioxide from the blood, or a kidney defect that leaves the blood deficient in bicarbonate. *See also* **alkalosis**.

■ **acid rain** Phenomenon caused by the pollution of the atmosphere with sulphur oxides and nitrogen oxides, which are produced largely by burning fossil fuels. The most common of these oxides are sulphur trioxide (SO_3), which combines readily with water to form sulphuric acid (H_2SO_4), and nitrogen dioxide (NO_2), which combines with water to form nitric acid (HNO_3). These acids are precipitated with snow and rain. [5.5.a]

acid value Measure of the amount of free **fatty acid** present in fat or oil. The value is given as the number of milligrams of potassium hydroxide required to neutralize the fatty acids in 1g of the substance being tested.

■ **acquired characteristics** Physical or behavioural characteristics that are acquired during an organism's lifetime (*e.g.* scars following wounds), and which cannot be genetically transferred to any offspring. *See also* **Lamarckism**.

■ **acquired immune deficiency syndrome** Virus disease more commonly known by its abbreviation **AIDS**.

acriflavine Orange crystalline organic compound used as an antiseptic. Alternative name: 2,8-diaminoacridine methochloride.

acromegaly Disorder of adults, involving overgrowth of bones, that usually results from continual production of growth hormone after the end of puberty. It may be caused by a tumour affecting the **pituitary**.

acrosome Cap-like structure at the tip of animal **sperm** cells. It contains **enzymes** that dissolve the membrane of the ovum during fertilization. [3.4.c]

ACTH Abbreviation of **adrenocorticotrophic hormone**.

actin Protein that, together with **myosin**, forms actomyosin in skeletal **muscle**. It exists in two forms: G-actin, which is globular, and its polymer, F-actin, which is fibrous.

Actinomycetes Group of mostly **saprophytic** bacteria that are characterized by branching multicellular filaments (hyphae).

Actinopterygii Subclass of the Osteichthyes, comprising all the ray-finned fish. [2.3.b] [2.4.a] [2.5.b]

■ **action potential** Rapid but transient change in the electrical potential across the membrane of an excitable cell (*e.g.* a nerve cell, in which it occurs during the passage of a nerve impulse). *See also* **all-or-none response**.

activated carbon Charcoal treated so as to be a particularly good absorbent of gases.

active site Part of an **enzyme** molecule to which its **substrate** is bound during **catalysis**.

■ **active transport** Energy-dependent movement of a dissolved substance across a cell membrane against a concentration gradient; *e.g.* taking up of nutrients by a plant's roots. This movement is normally in the opposite direction to that in which the substance would move by a passive diffusion process.

actomyosin Substance formed from the association of **actin** and **myosin** in skeletal **muscle**, upon which the contraction mechanism is based.

acupuncture System of alternate medicine that treats disorders or induces anaesthesia by inserting needles into the skin. The needles, which may be rotated or heated, are inserted along lines or meridians that are thought to influence various organs or parts of the body.

acute In medicine, describing a condition or disorder of sudden onset (as opposed to **chronic**).

■ **adaptation** *1.* Characteristic of an organism which improves that organism's chance of survival in its environment. *2.* Change in the behaviour of an organism in response to environmental conditions. *3.* Change in the sensitivity of a sensory mechanism after it has been exposed to a particular continuous stimulus. This allows the mechanism to adjust the sensitivity scale according to the level of the stimulus.

adaptive radiation Mechanism of **evolution** in which a single ancestor gives rise to a number of species that coexist but occupy different ecological niches. Alternative name: divergent evolution.

■ **addiction** Physical or psychological dependence on a drug, such as alcohol, tobacco, cocaine or heroin. Treatment of addiction is notoriously difficult, and stopping the drug produces unpleasant withdrawal symptoms. Psychiatric counselling may help. [3.4.b] [3.6.e]

■ **additive** Substance added to food to modify its colour, flavour or texture, or to preserve it. Some additives are natural substances, but many are synthetic compounds. A few have been implicated as possible causes of medical problems (such as allergy or hyperactivity) in sensitive persons.

adenine One of the component bases of **nucleic acids**. It is a **purine** derivative, and pairs with **thymine** in **DNA** and **uracil** in **RNA**. Alternative name: 6-aminopurine.

adenoids Pair of **lymph** glands, more prominent in children than in adults, that are located in and guard the back of the nasal cavity. Repeated infection of the adenoids causes them to become enlarged, and they may then be surgically removed (adenoidectomy). *See also* **tonsil**.

adenosine Compound consisting of the base **adenine** linked to the sugar **ribose**. Its phosphates are important energy carriers in biochemical processes (*see* next three articles).

■ **adenosine diphosphate** (ADP) Chemical involved with the biological transfer of chemical energy. Energy is released when it is formed from **adenosine triphosphate** (ATP). ADP is converted to ATP during **respiration** by addition of a phosphate group linked with an energy-rich bond.

adenosine monophosphate (AMP) Chemical involved with the biological transfer of chemical energy. Energy is released when it is formed from **adenosine triphosphate** (ATP). It is converted back to ATP during **respiration**.

■ **adenosine triphosphate** (ATP) Chemical that provides the energy for a large number of biological processes, including muscle contraction and the synthesis of many molecules. Chemically, it is an **adenosine** molecule with three attached phosphate groups. It gives off energy when it loses one or two of its phosphate groups (in a process known as **phosphorylation**) to become **adenosine diphosphate** or **adenosine monophosphate**, respectively. *See also* **electron transport**.

adenovirus One of a group of **viruses** that contain DNA and are found in a number of animals, including human beings. They cause infections of the respiratory tract, and may induce cancerous tumours.

■ **ADH** Abbreviation of **antidiuretic hormone**. *See also* **vasopressin**.

■ **adipose tissue** Tissue that contains cells specialized for the storage of fat. It gives insulation and acts as an energy reserve.

■ **ADP** Abbreviation of **adenosine diphosphate**.

adrenal cortex Outer part of the **adrenal gland**.

adrenal gland Endocrine gland that occurs in vertebrates (in mammals there are two glands, located on the kidneys). Each gland has two sections: the central medulla, which secretes the hormones **adrenaline** and **noradrenaline**; and the outer cortex, which secretes certain **steroid** hormones. Alternative name: suprarenal gland. [3.4.a] [3.5.e] [3.8.a]

■ **adrenalin Hormone** secreted by the medulla of the **adrenal gland** and at some nerve endings of the **sympathetic nervous system**. It is produced when the body prepares for violent physical action. Its effects include increased heartbeat, raised levels of sugar (glucose) in the blood and improved muscle action. Alternative names: adrenaline, epinephrine.

adrenergic Releasing or stimulated by **adrenalin** or **noradrenalin**. The term is applied to the **sympathetic nervous system**.

adrenocorticotrophic hormone (ACTH) **Protein** released by the anterior **pituitary**. It controls secretion from the cortex of the **adrenal gland**. Alternative name: corticotrophin.

■ **adventitious** Growing in an abnormal position, *e.g.* a root that develops from a stem.

aerobe Organism that respires aerobically (using oxygen). *See* **aerobic respiration**. [3.6.a]

■ **aerobic respiration** Process by which cells obtain energy from the **oxidation** of fuel molecules by **molecular oxygen** with the formation of carbon dioxide and water. This process yields more energy than **anaerobic respiration**. *See also* **glycolysis**; **Krebs cycle**. [3.6.a]

■ **aestivation** *1*. In zoology, summer dormancy or inactivity. It is characteristic of desert animals and others that survive very hot or dry periods by sleeping underground or in deep shade; *e.g.* lungfish burrow into mud at the beginning of the dry season and remain there until the rains come.

aetiology Study of the cause of disease. Alternative spelling: etiology. *See also* **epidemiology**.

■ **afferent** Leading towards (*e.g.* towards an organ of the body); the term is particularly used of various nerves and blood vessels. *See also* **efferent**.

affinity In biology, a similarity of structure or form between different species of plants or animals.

afforestation Planting of trees to create new forest, often to combat erosion.

aflatoxin One of the group of **toxins** produced by fungi of the genus *Aspergillus*. Animal feed (e.g. cereals, peanuts) contaminated by aflatoxins can cause outbreaks of disease in livestock. Aflatoxins are believed to be **carcinogenic** to human beings.

AFP Abbreviation of **alphafetoprotein**.

■ **afterbirth** Alternative name for the **placenta**. [3.4.c]

agamospermy Formation of plant seeds and embryos by asexual means (*see* **asexual reproduction**).

■ **agar** Complex **polysaccharide** obtained from seaweed, especially that of the genus *Gelidium*. It is commonly employed as a gelling agent in media used for growing **micro-organisms** and in food for human consumption. Alternative name: agar-agar.

■ **agglutination** Clumping or sticking together, *e.g.* of red blood cells from incompatible blood groups, or of bacteria. Agglutination is commonly used as an end point in immunological tests.

agglutinin **Antibody** that causes **agglutination**. The term commonly refers to such an antibody in a person's blood plasma that reacts with the red blood cells of another person's blood.

Agnatha Class of animals that includes the jawless fish (*e.g.* hagfish and lampreys).

agonist Substance (or sometimes an organ) that has an action that is complementary to the action of another substance (or organ) in an animal or a plant. The term is particularly used of drugs and muscles. *See also* **antagonist**.

agranulocytosis Blood disorder in which there is a deficiency of **granulocytes**, resulting in an impaired immune system.

agriculture Practice of cultivating the land to grow crops and rearing animals for food. *See* **horticulture**.

■ **AIDS** Abbreviation of acquired immune deficiency syndrome, a disease caused by the HIV (human immunodeficiency) virus, transmitted through the exchange of body fluids such as blood or semen. [3.4.b] [3.8.b]

■ **air** Mixture of gases that forms the Earth's atmosphere. Its composition varies slightly from place to place – particularly with regard to the amounts of carbon dioxide and water vapour it contains – but the average composition of dry air is (percentages by volume):

nitrogen 78.1%

oxygen 20.9%

argon 0.9%

other gases 0.1%

airbladder Alternative term for **swimbladder**.

■ **air sac** *1.* Cavity in the upper **thorax** of birds that is connected by passages to the lungs, and which increases the efficiency of breathing. *2.* Extension of the **trachea** in insects that increases the efficiency of oxygen uptake. *See also* **vesicle**.

alanine $CH_3C(NH_2)COOH$ **Amino acid** commonly found in proteins. Alternative name: 2-aminopropanoic acid.

■ **albinism** Pigment deficiency in animals or plants. In mammals, including human beings, it may affect the hair (which is white), skin and eyes. It is commonly caused by a

recessive gene which results in a lack of the enzyme that controls the synthesis of the dark pigment **melanin**.

■ **albumen** White of an egg.

■ **albumin** Soluble **protein** found in many animal fluids, most notably in egg white and **blood serum**.

■ **alcohol** Organic compound characterized by having a hydroxyl (–OH) group. The simplest alcohols are methanol (methyl alcohol, CH_3OH) and ethanol (ethyl alcohol, C_2H_5OH). Ethanol is the alcohol in alcoholic drinks such as beer, wine and spirits. [3.4.b] [3.6.e]

■ **alcoholism** Disorder caused by **addiction** to alcoholic drinks. It affects the sufferer's mental functions and leads to a deterioration of physical skills. The heart, liver and nerves may also be affected. If an alcoholic suddenly stops drinking, he or she experiences withdrawal symptoms, including anxiety, tremor and sometimes hallucinations.

aldose Type of **sugar** whose molecules contain an aldehyde (–CHO) group and one or more alcohol (–OH) groups.

aldosterone Hormone produced by the cortex of the **adrenal gland**, which affects the rate of carbohydrate **metabolism**. It also helps to control the **electrolyte** balance of the body by allowing the retention of sodium ions and the excretion of potassium ions.

■ **alga** Simple plant, which may be **unicellular** or **multicellular** (*e.g.* some seaweeds and the green slime in ponds). Algae contain a variety of photosynthetic pigments (hence brown, green and red seaweeds), and are present in many habitats; most are aquatic.

alginic acid $(C_6H_8O_6)_n$ Yellowish-white organic solid, a polymer of mannuronic acid in the **pyranose** ring form, that occurs in brown seaweeds. Even very dilute solutions of the acid are extremely viscous, and because of this property it has many industrial applications.

■ **alimentary canal** Tube in the body of animals along which food passes (moved by **peristalsis**), and in which the food is subjected to physical and chemical **digestion** and is absorbed. Alternative names: digestive tract, gut. [3.5.c]

■ **alkali** Substance that is either a soluble **base** or a solution of a base. Alkalis have a **pH** of more than 7 (pH = 7 is neutral, a pH of less than 7 is acidic), and react with **acids** to produce **salts** (and water).

■ **alkaline** Having the properties of an **alkali**.

■ **alkaloid** One of a group of nitrogen-containing organic compounds that are found in some plants. Many are toxic or medicinal (*e.g.* atropine, digitalis, heroin, morphine, quinine, strychnine).

alkalosis Condition in which the **pH** of the blood and body fluids rises above its normal level of 7.35. It may result from loss of acidic digestive juices through vomiting or from an excess intake of alkali (such as bicarbonate). *See also* **acidosis**.

allantois Membranous sac that develops in the **embryos** of reptiles, birds and mammals. It is involved in the storage of waste and the provision of food and oxygen. In reptiles and birds, it grows to surround the embryo in the shell; in mammals it becomes incorporated into the **placenta**.

■ **allele** One of the alternative states of a **gene**. In **diploid** cells each gene occurs in two forms; one may be genetically **dominant**, the other **recessive**. Alternative name: allelomorph. [4.8.a]

■ **allergy** Abnormal sensitivity of the body to a substance (known as an allergen). Contact with the allergen causes symptoms such as skin rashes, watery eyes, and sneezing. Hay fever is a widespread allergy to the pollen of certain plants. [3.8.b]

allogamy Fertilization of the **ovule** of a flowering plant that involves pollen from another flower (whether on the same plant or another plant).

■ **all-or-none response** Response of excitable tissue (*e.g.* some nerve cells) that occurs in full only at or above a certain level of stimulus, the threshold, but not at all below that level. *See also* **action potential**. [3.7.c]

alphafetoprotein (AFP) Protein that occurs in **amniotic fluid**, made in the liver of a foetus. High levels (detected by **amniocentesis**) may indicate a foetal abnormality (*e.g.* spina bifida).

■ **alternation of generations** Phenomenon that occurs in the life cycles of certain organisms (*e.g.* mosses, ferns and many coelenterates). A generation that reproduces sexually alternates with a generation that reproduces asexually. Consequently the life cycle is divided into **haploid** and **diploid** phases.

altruism Category of animal behaviour, especially common among social insects, in which older individuals tend to sacrifice themselves, losing their lives if necessary, in order that offspring and other younger individuals may survive or otherwise benefit.

■ **alveolus** Minute air sac in the vertebrate **lung**. There are vast numbers of alveoli, and most of the exchange of gases between air and blood takes place within them. *See also* **respiration**.

Alzheimer's disease Condition characterized by progressive degeneration of brain function, often referred to as premature senility. Disorders of speech and memory are the commonest symptoms. Recent research suggests that some cases of the disease may be linked to high levels of dissolved aluminium in drinking water. It was named after the German physician Alois Alzheimer (1864–1915).

ambergris Pale waxy solid with a strong smell which is believed to be produced inside the intestines of diseased sperm whales. Other than whaling, its only source is lumps washed ashore by the tide. It is used in the perfume industry as a fixative.

aminase One of a group of **enzymes** that can catalyse the **hydrolysis** of **amines**.

amination Transfer of an **amino group** ($-NH_2$) to a compound.

amine Member of a group of organic chemical substances in which one or more of the hydrogen atoms of **ammonia** (NH_3) have been replaced by a **hydrocarbon** group. In primary amines one, in secondary amines two and in tertiary amines three hydrogens have been so replaced.

■ **amino acid** The building blocks of **proteins**, amino acids are organic compounds that contain an acidic **carboxyl group** ($-COOH$) and a basic **amino group** ($-NH_2$). Twenty amino acids are commonly found in proteins. Those that can be synthesized by a particular organism are known as 'non-essential'; 'essential' amino acids must be obtained from the environment, usually from food.

amino group Chemical group with the general formula $-NRR'$, where R and R′ may be **hydrogen** atoms or organic

radicals; the commonest form is $-NH_2$. Compounds containing amino groups include **amines** and **amino acids**.

■ **ammonia** NH_3 Colourless pungent gas, which is very soluble in water (to form ammonium hydroxide, or ammonia solution, NH_4OH) and alcohol. It is formed naturally by the bacterial decomposition of proteins, purines and urea; made in the laboratory by the action of alkalis on ammonium salts; or synthesized commercially by fixation of nitrogen.

ammonia solution Alternative name for **ammonium hydroxide**.

ammonite Coiled-shelled **cephalopod** mollusc, now extinct, that was common in the seas of the Mesozoic era.

ammonium hydroxide NH_4OH **Alkali** made by dissolving **ammonia** in water, giving a solution that probably contains hydrates of ammonia. It is used for making soaps and fertilizers. 880 ammonia is a saturated aqueous solution of ammonia (density 0.88 g cm^{-3}). Alternative name: ammonia solution.

■ **amniocentesis** Test in which a sample of **amniotic fluid** is taken from the **amnion** surrounding a foetus. The cells that this contains are examined for foetal abnormalities, particularly hereditary disorders.

■ **amnion** Innermost membrane that envelops an **embryo** or **foetus** (in mammals, reptiles and birds) and encloses the fluid-filled amniotic cavity.

amniote Any of the higher vertebrates in which the **embryo** is surrounded by an **amnion** containing fluid (*i.e.* reptiles, birds, mammals). The development of the amnion, which permits gas exchange, was the evolutionary step that first enabled eggs to be laid on dry land.

■ **amniotic fluid** Liquid that occurs in the **amnion** surrounding a foetus.

■ **amoeba** Single-celled organism, a protozoan, that moves and feeds by the projection of **pseudopodia**.

■ **amoebocyte Cell** that demonstrates amoeboid movement. Such cells may be normally present in body fluids. *See also* **leucocyte**.

amorphous Without clear shape or structure.

AMP Abbreviation of **adenosine monophosphate**.

■ **amphetamine** Drug that stimulates the **central nervous system**. It was once much used for the treatment of depression and to lessen appetite, but is now seldom prescribed because of its addictive properties. [3.4.b] [3.6.e]

■ **Amphibia** Class of semi-aquatic **vertebrates** that evolved from fish. It includes frogs, toads, newts and salamanders. Amphibians are characterized by the possession of four five-toed legs and an inner ear. Eggs are laid in water, and fertilization is external. The larval stage is a free-swimming tadpole. [2.3.b] [2.4.a]

amphibian Member of the class **Amphibia**.

amphipod Member of the **Amphipoda**.

Amphipoda Order of small **crustaceans** that lack a **carapace** (*e.g.* sandhoppers, beach-hoppers and water lice). There are more than 4,000 species throughout the world, from Arctic seashores to the damp soils of tropical rainforests.

■ **amylase** Member of a group of **enzymes** that digest **starch** or **glycogen** to **dextrin**, **maltose** and **glucose**. Amylases are present in digestive juices and micro-organisms. Alternative name: diastase.

amyl nitrite $C_5H_{11}ONO$ Pale brown volatile liquid organic compound, often used in medicine to dilate the blood vessels of patients with some forms of heart disease (*e.g.* angina).

amylose Polysaccharide, a polymer of **glucose**, that occurs in **starch.**

amylum Alternative name for **starch**.

■ **anabolic steroid** Compound that is concerned with **anabolism**. Commonly used anabolic steroids are synthetic male **sex hormones** (androgens) which promote protein synthesis (hence their use by some athletes wishing to build up muscle). *See also* **steroid**. [3.4.b] [3.6.e]

■ **anabolism** Phase of **metabolism** that is concerned with the building up (or biosynthesis) of molecules. *See also* **catabolism**.

■ **anaemia** General term for any disorder characterized by abnormally low levels of **haemoglobin** in the blood. There are many different forms, which may be inherited or result from an infection or bodily disorder. A frequent cause is lack of iron in the diet. [3.4.b]

anaerobe Organism that respires anaerobically. See **anaerobic respiration**. [3.6.a]

■ **anaerobic respiration** Process by which organisms obtain energy from the breakdown of food molecules in the absence of oxygen (*e.g.* in plants **fermentation**, in which sugar is broken down to alcohol). In animals muscle cells respire anaerobically to form lactic acid. Both processes yield less energy than **aerobic respiration**. [3.6.a] [3.8.a]

■ **anaesthetic** Drug that induces overall insensibility (general anaesthetic) or loss of sensitivity in one area (local anaesthetic). [3.4.b]

■ **analgesic** Drug that relieves pain without causing loss of consciousness. [3.4.b] [3.6.e]

■ **analogous** In biology, describing structural features that have similar functions but which have developed completely independently in different plant or animal groups (*e.g.* the wings of birds and insects). *See also* **homologous**.

■ **anaphase** Stage in **mitosis** and **meiosis** (cell division) in which **chromosomes** migrate to opposite poles of the cell by means of the **spindle**.

anaphylaxis Sudden severe reaction (an **allergy**) to a drug or venom, requiring urgent medical treatment. Alternative name: anaphylactic shock.

■ **anastomosis** *1.* Natural vessel that connects two blood vessels. *2.* Artificially made connection between two body tubes (*e.g.* in the gut).

■ **anatomy** Study of the structural forms and minute structures of **plants** and **animals**.

androecium The male parts in a flower (the stamens) or a moss.

androgen Type of **hormone** that is associated with the development of male characteristics in vertebrates; a male sex hormone.

aneurin Alternative name for **thiamine** (vitamin B_1).

■ **angiosperm** Member of a major group of (flowering) plants, a section of the **Spermatophyta**, whose characteristics include the possession of flowers and the production of seeds contained in a fruit. [2.5.b]

■ **animal** Member of a large kingdom of organisms that feed **heterotrophically** on other organisms or organic matter.

Animals are usually capable of locomotion and movement from place to place and react quickly to stimuli (because they have sense organs and a nervous system). Locomotion and sensitivity are in proportion to their need to hunt for food. Animal cells have limited growth, no **chlorophyll**, and are surrounded by a cell membrane. [2.3.b] [2.4.a] [2.5.b]

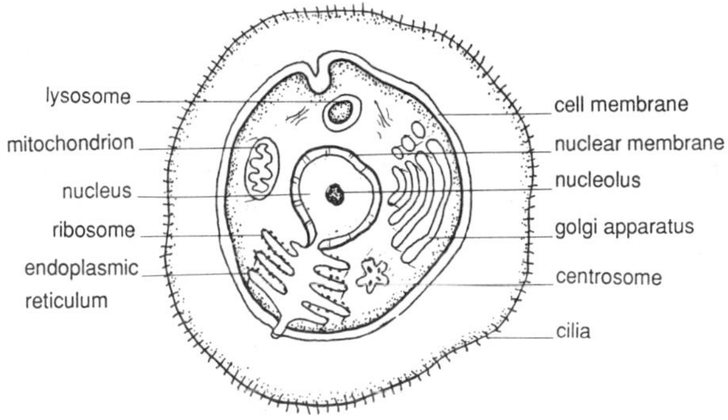

An animal cell

■ **animal starch** Alternative name for **glycogen**.

anisogamy **Fertilization** of **gametes** of slightly different types or sizes.

■ **annelid** Worm that is a member of the phylum **Annelida**. [2.3.b] [2.4.a] [2.5.b]

■ **Annelida** Phylum of invertebrates that consists of segmented worms (the true worms). The segmentation is both internal and external (earthworms and leeches are annelids). Annelid classes include **Oligochaeta** and **Polychaeta**.

■ **annual** Plant that completes its life cycle (from germination, through flowering and fruiting to death) in a single season. [2.3.c]

■ **annual ring** Yearly addition of **xylem** tissue to the stem of a woody plant, by means of which the age of the plant can be estimated. Alternative name: growth ring.

annulus *1.* In biology, any ring-shaped structure. *2.* Ring of tissue around the stem of the fruiting body of certain **fungi** after expansion of the **pileus**.

ANS Abbreviation of **autonomic nervous system**.

■ **antacid** Substance used medicinally to combat excess stomach acid (*e.g.* various compounds of **magnesium**). [3.4.b]

■ **antagonistic muscles** Muscles that work in pairs, one flexing (bending) a limb and the other straightening it.

■ **antenna** In biology, long thin jointed sensory structure in **arthropods** and some **fish**; a feeler.

■ **anterior** Near the head of an **animal**. *See also* **posterior**.

■ **anther Pollen**-producing part of a **stamen**. [3.4.a] [3.5.e]

antheridium Male **gamete**-producing organ in lower plants.

antherozoid Male sex cell (gamete) in an **antheridium**.

anthocyanin Member of a group of **flavonoid** pigments that give red, purple and blue colours to flowers and fruits.

Anthozoa Class of marine invertebrate animals that includes sea anemones, corals and sea-pens. Most anthozoans are sedentary, and all feed through a central mouth which is surrounded by one or more rings of tentacles.

anthrax Bacterial disease that affects many animals, including human beings, and which is frequently fatal. The disease-causing bacterium can form **spores**, and remain dormant in the soil or on stored animal hides for many years.

■ **anthropoid** Man-like. The term strictly refers to the members of the suborder Anthropoidea, which consists of monkeys, apes and man. [2.3.b] [2.4.a] [2.5.b]

anthropology Study of the human race. Physical anthropology is concerned with human evolution, social anthropology with behaviour.

■ **antibiotic** Member of a group of chemical substances that are by-products of **metabolism** in certain moulds and micro-organisms. Antibiotics cause the destruction of other micro-organisms, and are used as drugs to kill bacteria. *See also* **penicillin**. [3.4.b]

■ **antibody** Highly specific molecule produced by the **immune system** in response to the presence of an **antigen**, which it neutralizes. *See also* acquired **immunity**. [3.4.b] [3.8.b]

■ **anticoagulant** Chemical substance (*e.g.* coumarin, heparin, warfarin) that prevents blood from clotting. *See also* **coagulant**.

■ **antidiuretic hormone** (ADH) **Hormone** secreted from the posterior **pituitary** and synthesized in the **hypothalamus**. In **mammals** it stimulates the reabsorption of water in the kidney, and thus diminishes the volume of **urine** produced. *See also* **vasopressin**.

■ **antigen** Foreign substance that stimulates the production of **antibodies**. Most antigens are proteins (*e.g.* bacterial toxins), but they may be other **macromolecules** (*e.g.* bacteria, pollen, transplanted tissue). [3.8.b] [3.4.b]

■ **antihistamine** Chemical substance that inhibits the action of **histamine** by blocking its site of action. It may be used to treat an **allergy**. [3.8.b]

antioxidant Compound used to delay the **oxidation** of substances such as food by molecular oxygen. Most antioxidants are organic compounds; natural ones are found in vegetable oils and in some fruits (as ascorbic acid, or vitamin C).

■ **antiseptic** Substance that prevents sepsis by killing bacteria or preventing their growth. Alternative name: germicide.

antiserum Blood **serum** containing **antibodies**, used in vaccines to treat or prevent a disease or to combat animal venom (*e.g.* snakebite).

■ **antitoxin** Type of **antibody** against a toxoid produced in the body by a disease organism or by **vaccination**.

antrum *1.* Cavity in a bone (*e.g.* a sinus). *2.* Part of the stomach next to the pylorus.

■ **anus** Posterior opening of the **alimentary canal**, through which the undigested residue of digestion is passed. [3.4.a] [3.5.e] [3.8.a]

■ **aorta** Principal **artery** that takes oxygenated blood from the heart to all parts of the body other than the lungs. [3.4.a] [3.5.e] [3.8.a] [3.8.c]

Apgar score Method of assessing the health of a newborn baby by assigning 0 to 2 points to the quality of heartbeat, breathing, muscle tone, skin colour and reflexes. It was named after the American physician Virginia Apgar (1909–).

apical growth Alternative name for **primary growth**.

■ **apical meristem** Zone of cell division in **vascular plants**. It consists of a group of cells at the tip of the stem or root from which all new tissues of the plant are formed.

apoenzyme Inactive **enzyme** that consists of a protein and a non-protein portion. It needs a **coenzyme** in order to function. *See also* **holoenzyme**.

apoplast Parts of a plant that do not consist of living tissue. These are **cell walls, xylem** and spaces between cells.

aposematic coloration Alternative name for **warning coloration**.

■ **appendix** Vestigial outgrowth of the **caecum** in some mammals. In human beings its full name is vermiform appendix.

■ **aqueous** Dissolved in water, or chiefly consisting of water.

■ **aqueous humour** Liquid between the **lens** and **cornea** of the **eye**. [15.6.b]

■ **aqueous solution** Solution in which the **solvent** is water.

arabinose $C_5H_{10}O_5$ Crystalline **pentose sugar** derived from plant **polysaccharides** (such as gums), sometimes used as a culture medium in bacteriology.

■ **arachnid** Member of the class **Arachnida**; *e.g.* a spider.

■ **Arachnida** Class of mainly terrestrial **arthropods**. Arachnids have a combined head and **thorax** (called a prosoma or cephalothorax) and an **abdomen**. The prosoma bears four pairs of legs, but no wings or antennae. Members of the class include mites, ticks, spiders and scorpions. [2.3.b] [2.4.a] [2.5.b]

arachnoid membrane One of the three thin membranes that cover the **brain**, the other two being the pia mater and the dura mater.

archegonium Spore-producing tissue from which the spore mother cells are formed in lower plants, such as ferns.

arenaceous In biology, describing plants that grow best in sandy soils, and animals that live mainly in sand.

■ **areola** *1.* Dark skin on the breast surrounding a nipple. *2.* Part of the iris of the **eye**, bordering the pupil. [3.4.a] [3.5.e] [3.8.a]

areolar tissue Connective tissue made up of cells separated by bundles of fibres embedded in **mucin**.

arginine $C_6H_{14}N_4O_2$ Colourless crystalline **essential amino acid** of the alpha-ketoglutaric acid family.

arrow-worm One of a group of finned worm-like animals that make up the phylum **Chaetognatha**.

■ **arteriole** Small **artery**.

■ **artery** **Blood vessel** that carries oxygenated blood from the heart to other tissues. An exception is the pulmonary artery, which carries deoxygenated blood to the lungs. [3.4.a] [3.5.e] [3.8.a] [3.8.c]

■ **arthropod** Member of the phylum **Arthropoda**. [2.3.b] [2.4.a] [2.5.b]

■ **Arthropoda** Phylum of invertebrates, the largest phylum in the animal kingdom, that consists of animals with jointed appendages, a well-defined head and usually a hard **exoskeleton** made of **chitin**. Arthropods have representatives in every habitat and include arachnids, crustaceans and insects.

■ **artificial insemination** *1.* Artificial implantation of **semen** containing sperm into the female **cervix**. *2.* Transfer of a fertilized **ovum** from the reproductive tract of one female to that of a host mother. [4.9.c]

■ **artificial selection** Causing the production of offspring that are most commercially useful by choosing the parents, in both plants and animals. Alternative name: selective breeding. [4.9.a] [4.9.b] [4.9.c]

Artiodactyla Order of mammals that comprises the even-toed (usually two toes) **ungulates**. Artiodactyls include antelopes, bison, buffaloes, camels, cattle, deer, giraffes, goats, pigs and sheep. All are **herbivores** and many are **ruminants**. *See also* **Perissodactyla**.

Ascomycetes Important class of **fungi** in which the **spore**-producing body is an **ascus**. It includes morels, truffles, the fungal part of most **lichens**, and many **yeasts**. Alternative name: Ascomycotina.

Ascomycotina Alternative name for **Ascomycetes**.

■ **ascorbic acid** $C_6H_8O_6$ White crystalline water-soluble **vitamin** found in many plant materials, particularly fresh fruit and vegetables. It is a natural **antioxidant**. Alternative name: vitamin C.

ascus Cell in fungi in which haploid **spores** are formed. *See* **Ascomycetes**.

■ **aseptic** Free from **pathogenic** micro-organisms (particularly bacteria).

■ **asexual reproduction** Reproduction that does not involve **gametes** or **fertilization**. There is a single parent, and all offspring are genetically identical (*see* **clone**). *See also* **vegetative propagation**. [3.4.c] [4.9.c]

aspartame Artificial sweetener (a dipeptide) that is 200 times as sweet as ordinary sugar (sucrose), but does not have the bitter aftertaste characteristic of **saccharin**.

■ **aspirin** $CH_3COO.C_6H_4COOH$ Drug that is commonly used as an **analgesic**, antipyretic (to reduce fever) and anti-inflammatory. Alternative name: acetylsalicylic acid. [3.4.b]

■ **assimilation** *1.* General term for the process by which food is made and used by **plants**. *2.* Process of turning food into body substances after it has been digested (*e.g.* in animals excess glucose is turned into glycogen or fat, amino acids are made into proteins).

aster phase Alternative name for **metaphase**.

■ **astigmatism** Vision defect caused by irregular curvature of the lens of the **eye**, so that light does not focus properly. [15.6.c]

■ **atlas** In anatomy, the first (cervical) vertebra, which joins the skull to the spine and articulates with the **axis**, allowing nodding movements of the head. *See also* **axis**.

■ **ATP** Abbreviation of **adenosine triphosphate**.

■ **atrium** One of the thin-walled upper chambers of the **heart**. Alternative name: auricle.

■ **atrophy** Wasting away of a **cell** or **organ** of the body.

atropine White crystalline poisonous **alkaloid** that occurs in deadly nightshade. Alternative name: belladonna.

■ **auditory** Relating to the **ear** or hearing.

■ **auditory canal** Tube that leads from the outer **ear** to the ear drum (tympanum). [3.4.a] [3.5.e] [3.8.a]

■ **auditory nerve** Nerve that carries impulses concerned with hearing from the inner **ear** to the brain. [3.4.a] [3.5.e] [3.8.a]

■ **auricle** Alternative name for **atrium**.

autism Brain disorder that develops in infancy and which is characterized by extreme learning difficulties and a lack of responsiveness to other people.

autocatalysis Catalytic reaction that is started by the products of a reaction that was itself catalytic. *See* **catalyst**.

■ **autoclave** Airtight container that heats and sometimes agitates its contents under high-pressure steam. Autoclaves are used for sterilizing surgical instruments, industrial processing and in biotechnology.

autogamy Uniting of **gametes** produced by the same **cell**. Alternative name: **self-fertilization.**

■ **autolysis** Breakdown of the contents of an animal or plant cell by the action of **enzymes** produced within that cell.

■ **autonomic nervous system** (ANS) In vertebrates, the part of the **nervous system** that is not under voluntary control and carries nerve impulses to the **smooth muscles**, glands, heart and other organs. It is subdivided into the **parasympathetic** and **sympathetic nervous systems**.

auto-oxidation **Oxidation** caused by the unaided **atmosphere**.

autoradiography Technique for photographing a specimen by injecting it with radioactive material so that it produces its own image on a photographic film or plate.

autosome Any **chromosome** other than one of the **sex chromosomes**.

autotroph Organism that lives using **autotrophism**.

■ **autotrophism** In bacteria and green plants, the ability to build up food materials from simple substances, *e.g.* by **photosynthesis**. *See also* **heterotrophism**.

■ **auxin** Any substance, including various hormones, that controls the growth of plants. [3.9.b]

■ **Aves** Class of vertebrates made up of birds. [2.3.b]

■ **axis** *1.* In biology, central line of symmetry of an organism. *2.* In anatomy, the second (cervical) vertebra, which articulates with the **atlas** and allows the head to turn from side to side. *3.* In constructing a graph, the vertical or horizontal line calibrated in numbers or units. [1.3.f] [1.4.h]

■ **axon** Thread-like outgrowth of a **nerve cell** (neurone), the main function of which is to carry impulses towards the cell body. Sensory neurones have long axons and short dendrons; motor neurones have short axons and long dendrons. *See also* **dendron**. [3.7.c]

B

Bacillariophyta Class of **algae** that are characterized by possession of cell walls impregnated with **silica**. Composed of two halves, they are microscopic, unicellular marine or freshwater organisms. Alternative name: diatoms.

■ **bacillus** *1.* Descriptive term for any rod-shaped **bacterium**. *2.* Specifically, *Bacillus* is a genus of spore-producing bacteria.

■ **backbone** Alternative name for **vertebral column**. [3.3.a]

■ **background radiation Radiation** from natural sources, including outer space (cosmic radiation) and radioactive substances on Earth (*e.g.* in igneous rocks such as granite).

■ **bacteria** Plural of **bacterium**.

■ **bactericide** Substance that can kill **bacteria**.

■ **bacteriology** Study of **bacteria**, their effects on organisms and their uses in agriculture and industry (*e.g.* in **biotechnology**).

bacteriophage Virus that infects **bacteria**. When inside a cell it replicates using its host's **enzymes**; the release of new viruses may disintegrate the cell. Bacteriophages have been used extensively in research on **genes**. Alternative name: phage.

bacteriostatic Substance that inhibits the growth of **bacteria** without killing them. *See also* **bactericide**.

■ **bacterium** Member of a group of microscopic organisms distinct from the plant and animal kingdoms. The shape of bacteria varies according to species, *e.g.* rod-shaped (bacillus), spherical (coccus), spiral (spirillum or spirochaete). Multiplication is fast, usually by fission, every 20 minutes in favourable conditions. Between them, the various species can utilize almost any type of organic molecule as food and inhabit almost any environment; there are those that can even use inorganic elements such as sulphur. The activities of some bacteria are of great significance to man, *e.g.* **fixation of nitrogen** in certain plants, as agents of decay, and their medical importance as disease-causing agents (pathogens).

■ **balance** Sense supplied by organs within the semicircular canals of the **inner ear**. [3.7.c]

■ **balanced diet** Food that contains nutrients – fat, carbohydrate, protein, vitamins, mineral salts, water and roughage (fibre) – in the correct proportions for good health.

baleen Fibrous material that hangs in plates inside the mouths of certain species of whales, where it acts to filter plankton (their food) from sea-water. Alternative name: whalebone.

■ **barbiturate** Sedative and hypnotic drug derived from barbituric acid, formerly used in sleeping pills but now generally employed only as a fast-acting anaesthetic and to treat epilepsy. [3.4.b] [3.8.b]

bar chart Graph in which data is presented as bars or blocks ranged along an axis. *See also* **histogram**.

■ **bark** Outer surface of plant **stems** and **roots** which have undergone secondary growth or thickening. It results from the activity of cork **cambium** and gives the plant a protective outer surface of dead **cells**. [3.5.e] [3.6.b]

Barr body Condensed **X-chromosome**, seen in cells of female mammals, due to one or other of the two X-chromosomes in each cell being inactivated. It was named after the Canadian biologist M. Barr (1908 –).

basal ganglion Region of grey matter within the white matter that forms the inner part of the cerebral hemispheres of the **brain**. Alternative name: basal nucleus.

■ **basal metabolic rate** (BMR) Minimum amount of energy on which the body can survive, measured by oxygen consumption and expressed in kilojoules per unit body surface. [3.6.a]

base *1*. In biology, one of the **nucleotides** of **DNA** or **RNA** (*i.e.* adenine, cytosine, guanine, thymine or uracil). *2*. In chemistry, a member of a class of chemical compounds whose aqueous solutions contain OH^- **ions**. A base neutralizes an **acid** to form a **salt** (*see also* **alkali**).

base code Sequence of bases on a strand of **DNA** that determines the type of information carried by a **gene**. [4.10.b]

base pairing Specific pairing between complementary **nucleotides** in double-stranded **DNA** or **RNA** by **hydrogen bonding**; *e.g.* in DNA, guanine pairs with cytosine and adenine pairs with thymine. *See also* **purine**; **pyrimidine**.

Basidiomycetes Major class of **fungi** that includes rusts and smuts, in which the **spores** form in a specialized club-shaped cell known as the basidium which develops inside the fruiting body. Alternative name: Basidiomycotina.

basidium *See* **Basidiomycetes**.

■ **BCG** Abbreviation of bacillus Calmette-Guérin, a **vaccine** used against tuberculosis. It was named after two French bacteriologists, Albert Calmette (1863–1933) and Camille Guérin. [3.8.b]

■ **beetle** Member of the **Coleoptera** order of insects, which typically have hard wing cases over membranous flight wings and biting mouthparts in the adult. The larvae may be active predators or sluggish grubs. [2.3.b] [2.4.a] [2.5.b]

■ **beet sugar** Alternative name for **sucrose**.

behaviourism One branch of psychology, which concentrates exclusively on observable actions and takes no account of mental events (thoughts, emotions, etc.) that cannot be experienced by the observer (the psychologist).

■ **Benedict's test** Food test used to detect the presence of a **reducing sugar** by the addition of a solution containing sodium carbonate, sodium citrate, potassium thiocyanate, copper sulphate and potassium ferrocyanide. A change in colour from blue to red or yellow on boiling indicates a positive result. It was named after the American chemist S. Benedict (1884–1936). *See also* **Fehling's test**. [3.10.a]

■ **benign** In medicine, describing a **tumour** that does not spread or destroy the tissue in which it is located; not life-threatening. *See also* **malignant**.

benthos Organisms that live on the bottom of a sea or lake.

benzodiazepine Member of a group of drugs used as antidepressants and anticonvulsants.

benzpyrene Cyclic organic compound, found in coal-tar and tobacco smoke, which has strong carcinogenic (cancer-forming) properties.

■ **berry** Fruit consisting of a fleshy **pericarp** containing seeds; *e.g.* cucumber, grape, lemon, melon, tomato (the date is an unusual berry in having only one seed). Botanically, blackberries, raspberries etc. are not true berries. [3.4.d]

beta-blocker Drug that slows the heartbeat, used to treat hypertension (high blood pressure). [3.4.b]

■ **biceps** Flexor muscle in the upper arm that has two points of origin and one insertion. It is one of a pair of **antagonistic muscles**, the triceps being the other (extensor) muscle.

■ **biennial Plant** whose life-cycle lasts for two years. Flowering and seed production occur in the second year of life. [2.3.c]

■ **bilateral symmetry** Type of symmetry in which a shape is symmetrical about a single plane (each half being a mirror image of the other). *E.g.* most vertebrates are bilaterally symmetrical. *See also* **radial symmetry**.

■ **bile** Alkaline mixture of substances produced by the liver (from waste products when blood cells are broken down) and stored in the **gall bladder**, which passes it to the **duodenum**, where it emulsifies fats (preparing them for digestion) and neutralizes acid. Its (yellowish) colour is due to bile pigments (*e.g.* **bilirubin**). [3.4.a] [3.5.e] [3.8.a]

■ **bile duct** Tube that carries **bile** from the gall bladder to the **duodenum**. [3.4.a] [3.5.e] [3.8.a]

■ **bilharzia** Disease that affects human beings and domestic animals in some subtropical regions, which results from an infestation by one of the parasitic blood flukes belonging to the genus *Schistosoma*. The larvae of the flukes develop inside freshwater snails and become free-swimming organisms which can attach themselves to a wading mammal, penetrating the skin and entering the bloodstream. Alternative name: schistosomiasis. [3.8.b]

biliary To do with **bile** or the **gall bladder**.

bilirubin Major pigment in **bile**, formed in the liver by the breakdown of **haemoglobin**. Its accumulation causes the symptom **jaundice**.

■ **binary fission** Method of reproduction employed by many single-celled organisms, in which the so-called mother cell divides in half (by **mitosis**), forming two identical, but independent, daughter cells. It is a type of **asexual reproduction**. [3.2.a] [3.5.d]

■ **binomial nomenclature** System by which organisms are identified by two Latin or pseudo-Latin names. The first is the name of the **genus** (generic name), the second is the name of the **species** (specific name), *e.g. Homo sapiens*. Alternative term: Linnaean system (after the Swedish botanist Carolus Linnaeus, 1707–78). *See also* **classification**.

bioassay Method for quantitatively determining the concentration of a substance by its effect on living organisms, *e.g.* its effect on the growth of **bacteria**.

biochemical oxygen demand (BOD) Oxygen-consuming property of natural water because of the organisms that live in it.

■ **biochemistry** Study of the chemistry of living organisms.

■ **biodegradation** Breakdown or decay of substances by the action of living organisms, especially **saprophytic** bacteria and fungi. Through biodegradation, organic matter is recycled. [5.5.b]

■ **bioenergetics** Study of the transfer and utilization of energy in living systems. *See also* **adenosine triphosphate**. [3.6.a]

■ **bioengineering** Application of engineering science and technology to living systems. [4.10.d]

biofeedback Method of controlling a bodily process (*e.g.* heartbeat) that is not normally subject to voluntary control by making the person concerned aware of measurements from instruments monitoring that process.

■ **biogenesis** Theory that living organisms may originate only from other living organisms, as opposed to the theory of **spontaneous generation**. *See also* **abiogenesis**.

■ **biological clock** Hypothetical mechanism in plants and animals that controls periodic changes in internal functions and behaviour independently of environment, *e.g.* diurnal rhythms and hibernation patterns.

■ **biology** Study of living organisms and their relation to the non-living environment. Its two major divisions are botany (the study of plants) and zoology (the study of animals).

■ **bioluminescence** Emission of visible light by living organisms, *e.g.* certain bacteria, fungi, fish and insects. It is produced by **enzyme** reactions in which chemical energy is converted to light.

■ **biomass** Total mass of living matter in a given environment or **food chain** level. [2.7.b] [5.2.a]

biome Subdivision of the living planet (the ecosphere), the largest area which ecology can deal with conveniently; an ecological region broadly equivalent to one of the climatic regions, *e.g.* rainforest or desert.

biometry Application of mathematical and statistical methods to the study of living organisms.

■ **bionic** Describing an artificial device or system that has the properties of a living one.

■ **biophysics** Use of ideas and methods of physics in the study of living organisms and processes.

■ **biopsy** Small sample of cells or tissue removed from a living subject for laboratory examination (usually as an aid to diagnosis).

■ **biorhythm** One of the cyclic pattern of 'highs' and 'lows' which some people believe govern each person's emotional, physical and intellectual behaviour. Most scientists are sceptical of the concept because it is based on the individual's subjective perceptions of these qualities. *See also* **circadian rhythm**.

■ **biosphere** Region of the Earth and its **atmosphere** which may be inhabited by living organisms. [5.7.a]

biosynthesis Formation of the major molecular components of cells, *e.g.* **proteins**, from simple components.

■ **biotechnology** Utilization of living organisms for the production of useful substances or processes, *e.g.* in **fermentation** and milk production.

biotin Coenzyme that is involved in the transfer of **carbonyl groups** in biochemical reactions, such as the metabolism of fats; one of the B vitamins.

■ **bird** Vertebrate animal of the class **Aves**, characterized by having feathers, laying hard-shelled eggs and (usually) the power of flight. [2.3.b] [2.4.a] [2.5.b]

■ **birth** Process by which young (*e.g.*of a mammal) separates from its mother. Alternative name: parturition. [3.4.c]

bisexual Having both male and female characteristics.

■ **biuret test** Test used to detect **peptides** and **proteins** in solution by treatment with **copper sulphate** and **alkali** to give a purple colour.

bivalve Animal of the class **Bivalvia**.

Bivalvia Class of **molluscs** comprising shellfish with a pair of hinged shells; *e.g.* clams, cockles, mussels, oysters and scallops. Alternative names: Lamellibranchiata, Pelecypoda.

■ **bladder** *1.* In lower plants, *e.g.* bladderwort, a modified leaf that catches small aquatic animals. *2.* In animals, a membranous sac containing gas or fluid, *e.g.* the air-filled swim bladder of bony fish, and the gall bladder and urinary bladder of mammals. *See also* **vesicle**.

blastocyst Early stage in the development of a mammalian **embryo**, formed by successive cleavages of the fertilized ovum. It consists of a solid ball of cells surrounded by a hollow sphere of cells, and goes on to form the **gastrula**. Alternative name: blastula.

blastula Hollow sphere composed of a single layer of **cells** produced by cleavage of a fertilized **ovum** in **animals**. In mammals, it is also called a blastocyst.

■ **blind spot** Area on the **retina** of the vertebrate **eye** which, because it is at the point of entry of the optic nerve, is without light-sensitive cells and is thus blind. [15.6.b]

blink microscope Instrument for comparing two very similar photographs, *e.g.* of bacteria. The photographs are viewed side by side, one with each eye, and are rapidly concealed and uncovered. The brain's attempts to superimpose the two images reveals any slight differences between them. Alternative name: blink comparator.

■ **blood** Fluid in the bodies of animals that circulates and transports oxygen and nutrients to cells, and carries waste products from them to the organs of excretion. It also transports hormones and the products of digestion. Essential for maintaining uniform temperature in warm-blooded animals, blood is made up mainly of **erythrocytes, leucocytes, platelets**, water and **proteins**. *See also* **haemocyanin**; **haemoglobin**. [3.5.c] [3.5.e] [3.7.b]

■ **blood cells** *See* **erythrocyte**; **leucocyte**; **lymphocyte**. [3.4.a] [3.5.e] [3.8.a]

blood clot Coagulated blood which may help wound healing or be the cause of blocking an artery (possibly leading to an infarction or stroke). Alternative name: thrombus. [3.4.a] [3.5.e] [3.8.a]

■ **blood group** *See* **ABO blood groups**. [3.4.a] [3.5.e] [3.8.a]

■ **blood plasma** Straw-coloured fluid part of **blood** in which blood cells are suspended. Its importance is for transporting blood cells, plasma **proteins, urea,** sugars, mineral salts, hormones and carbon dioxide as bicarbonates in solution. [3.4.a] [3.5.e] [3.8.a]

■ **blood poisoning** Alternative name for **septicaemia**. [3.4.a] [3.5.e] [3.8.a]

■ **blood pressure** Pressure of blood flowing in the arteries. It varies between the higher value of the systolic pressure (when the heart's ventricles are contracting and forcing blood out of the heart) and the lower value of the diastolic pressure (when the heart is filling with blood). It is affected by exercise, emotion, certain illnesses and various drugs. *See also* **heartbeat**. [3.4.a] [3.5.e] [3.8.a]

■ **blood serum** Fluid part of **blood** from which all blood cells and **fibrin** have been removed. It may contain **antibodies**, and such serums are used as **vaccines**. [3.4.a] [3.5.e] [3.8.a]

■ **blood sugar** The energy-generating sugar **glucose**, whose level in the blood is controlled by the hormones **insulin** and **adrenalin**. [3.4.a] [3.5.e] [3.8.a]

■ **blood vascular system** System consisting of the **heart, arteries, veins** and **capillaries**. The heart acts as a central muscular pump which propels oxygenated **blood** from the lungs along arteries to the tissues; deoxygenated blood is carried along veins back to the heart and lungs. [3.4.a] [3.5.e] [3.8.a]

■ **blood vessel** An **artery**, **vein** or **capillary**. Many of the larger blood vessels have muscular walls, whose contraction and relaxation aids blood flow. [3.4.a] [3.5.e] [3.8.a]

bloom Alternative name for **flower**.

■ **blue-green alga** Alternative name for a member of the **Cyanophyta**.

■ **BMR** Abbreviation of **basal metabolic rate**.

■ **body cavity** In **triploblastic** animals, an internal cavity bounded by the body wall. It contains the **viscera**.

■ **bone** Skeletal substance of **vertebrates**. It consists of **cells** (osteocytes) distributed in a matrix of **collagen** fibres impregnated with a complex salt (bone salt), mainly **calcium phosphate**, for hardness. Cells are connected by fine channels that permeate the matrix. Larger channels contain **blood vessels** and **nerves**. Some bones are hollow and contain **bone marrow**. There are about 206 bones in an adult human skeleton. [3.3.a]

■ **bone marrow** Soft tissue that fills the centre of some **bones**. It is responsible for the manufacture of the majority of **blood** components, including **erythrocytes** (red blood cells) and **lymphocytes** (cells involved in the immune response). [3.3.a]

bony fish Alternative name for **Osteichthyes**.

■ **botany** Study of **plants** and plant life. [2.1.a]

■ **botulism** Severe food poisoning, an often fatal disorder caused by eating food contaminated with a **toxin** produced by the bacterium *Clostridium botulinum* (which is destroyed by adequate cooking). Botulinus toxin is one of the most potent poisons known. [3.8.b]

■ **bowel** Alternative name for **intestine**. [3.4.a] [3.5.e] [3.8.a]

■ **Bowman's capsule** Dense ball of capillary blood vessels that cover the closed end of every **nephron** in the **kidney.** From it leads the uriniferous tubule. It was named after the British physician William Bowman (1816–92). [3.8.a] [3.8.c]

■ **Brachiopoda** Phylum of marine shellfish that evolved before the **molluscs**. Many species are now extinct, but other brachiopods remain as some of the oldest 'living fossils'.

■ **bract** Leaf-like structure in whose axil a **flower** grows.

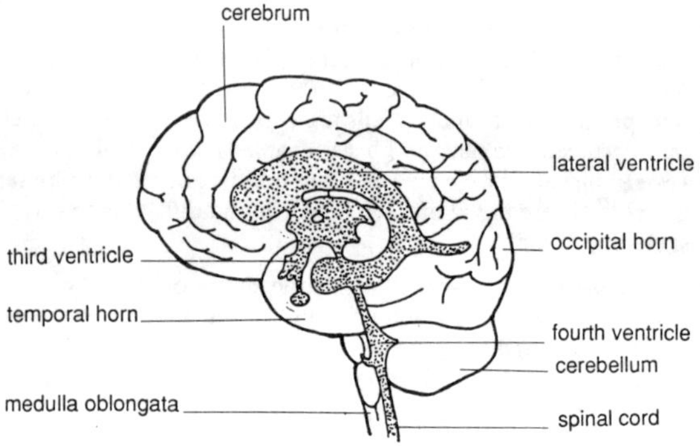

The human brain

■ **brain** Principal collection of **nerve cells** that form the anterior part of the central **nervous system**, consisting (in mammals) of a moist pinkish-grey mass protected by the bones of the cranium (skull). It receives, mostly via spinal nerves from the spinal cord but also via cranial nerves from organs in the head, sensory information through **afferent** sensory **neurones** carrying the impulses from sense organs, and it sends out instructions along **efferent** (motor) neurones to the effector organs (*e.g.* muscles). It is also the centre of intellect and memory so that behaviour can be based on past experience, and is responsible for the co-ordination of the whole body. *See also* **cerebellum**; **cerebrum**; **cortex**. [3.5.e]

■ **brain stem** Part of the brain at the end of the spinal cord, consisting of the midbrain, **medulla oblongata** and **pons**.

■ **brainwave** Pattern of electrical activity in the brain as revealed by an electroencephalograph. Alpha waves correspond to wakefulness with eyes closed, beta waves to wakefulness with eyes open and delta waves to deep sleep.

■ **breastbone** Alternative name for **sternum**.

■ **breathing** The physical actions of inhalation, gaseous exchange and exhalation. *See* **respiration**. [3.3.a] [3.6.a]

■ **breed** Artificial subdivision of a species resulting from its domestication and selective breeding by human beings, to produce food animals (*e.g.* cattle, pigs, sheep), working animals (*e.g.* horses, dogs) or pets (*e.g.* cats).

brine Concentrated solution of common salt (**sodium chloride**).

Broca's area Speech centre of the **brain**. It was named after the French surgeon Paul Broca (1824–80).

■ **bronchiole** Terminal air-conducting tube (1mm in diameter) of mammalian **lungs**, arising from secondary subdivision of a **bronchus** and terminating in **alveoli**. [3.6.a]

bronchitis Disorder caused by inflammation of the **bronchi** of the lungs resulting from infection or prolonged exposure to irritant chemicals such as tobacco smoke or pollutants. The airways to the lungs become narrowed and clogged with mucus, causing breathlessness, chest pain and a persistent cough. [3.5.d] [3.6.a]

■ **bronchus** One of the two air-carrying divisions of the **trachea** (windpipe) into the **lungs**. The bronchi become inflamed in the disorder **bronchitis**. [3.6.a]

Bryophyta Division of the plant kingdom that consists principally of the liverworts and mosses. In evolutionary terms, bryophytes are primitive plants which exhibit **alternation of generations**.

bryophyte Member of the plant division **Bryophyta**.

■ **bud** *1.* In plants, an immature shoot that will bear leaves or flowers. *2.* In fungi and simple animals, a small outgrowth from a parent organism capable of detaching itself and living an independent existence. [3.2.a]

■ **budding** *1.* Formation of **buds** from specialized cells on a plant shoot. *2.* **Grafting** of a bud onto a plant. *3.* Method of **asexual reproduction** employed by **yeasts** and simple animal organisms such as hydra; in plants it is termed gemmation. [3.4.c] [3.8.a] [3.9.b] [4.9.c]

■ **bug** Plant-sucking or blood-sucking insect of the order Hemiptera. The term is also used (incorrectly) for beetles, fleas and other insects.

■ **bulb** In botany, an underground perennating organ developed by some plants. It is usually formed from fleshy leaf bases, and has adventitious roots growing from a small piece of stem at its lower surface. Primarily a storage organ, it may also play a part in **vegetative propagation**. [3.4.d]

C

■ **caecum** Pouch or pocket, such as one at the junction of the small and large intestines from which hangs the **appendix.**

■ **caffeine** White **alkaloid** with a bitter taste, obtained from coffee beans, tea leaves, kola nuts, or by chemical synthesis. It is a diuretic and a stimulant to the **central nervous system**.

calcaneus Heel-bone; the major bone of the foot. Alternative name: calcaneum.

calciferol Fat-soluble **vitamin** formed in the skin by the action of sunlight, which controls levels of calcium in the blood. Alternative name: vitamin D.

■ **calcification** Depositing of calcium salts in body tissue, a normal process in the formation of **bones** and **teeth** but abnormal in the formation of **calculi** ('stones'). *See also* **cartilage**.

calcitonin Hormone secreted in vertebrates that controls the release of calcium from bone. In mammals it is secreted by the **thyroid gland**. Alternative name: thyrocalcitonin.

calculus Abnormal hard accretion (a 'stone') of calcium salts and other compounds that may form in the kidneys and urinary tract, gall bladder, bile ducts or salivary glands.

■ **calorie** (with a small *c*) Amount of heat required to raise the temperature of 1g of water by 1°C at one atmosphere pressure; equal to 4.184 joules. *See also* **Calorie**.

■ **Calorie** (with a capital C) Amount of heat required to raise the temperature of 1kg of water by 1°C at one atmosphere pressure; equal to 4.2 kilojoules. It is used as a unit of energy of food (when it is sometimes spelled with a small *c*). Alternative name: kilocalorie, large calorie. *See also* **calorie**.

■ **calyx** Outermost **whorl** of a flower composed of leaf-like **sepals**.

■ **cambium** Cellular tissue in which **secondary growth** occurs in the **stem** and **root** of a plant.

camphor Naturally occurring organic compound with a penetrating aromatic odour. Alternative names: gum camphor, 2-camphanone.

■ **cancer** Disorder that results when body cells undergo unrestrained division (because of a breakdown in the normal control of cellular processes) to form a malignant **tumour**. Cancer may spread by **metastasis,** when tumour cells spread from their original position and invade other tissues.

■ **cane-sugar** Alternative name for **sucrose**.

■ **canine tooth** Pointed tooth in mammals (except most rodents and ungulates – *i.e.* herbivores), adapted for stabbing and tearing food. There are two canines in each jaw. Alternative name: eye-tooth.

■ **capillarity** Means by which water and other fluids rises up very narrow tubes (because of the cohesive nature of water molecules). Examples include the movement of water between soil particles or up **xylem** vessels in a plant. Alternative name: capillary action.

■ **capillary** In biology, finest vessel of the **blood vascular system** in vertebrates. Large numbers of capillaries are present in **tissues**. Their walls are composed of a single layer of cells through which exchange of substances, *e.g.* **oxygen**, occurs between the tissues and **blood**. [3.5.e]

capillary action Alternative term for **capillarity**.

■ **capillary tube** Very fine tube up which water or other fluid rises by **capillarity**.

■ **capsule** *1.* In certain **bacteria**, a gelatinous extracellular envelope, which in some cases protects the cell. *2.* In mosses, structure in which spores are formed. *3.* In flowering plants, dry fruit that may liberate seeds in various ways (*e.g.* poppy). *4.* In anatomy, sheath of membrane that surrounds an organ or area of tissue; *e.g.* the synovial capsule that surrounds the moving parts of joints. [3.4.d]

■ **carapace** Hard shield that covers the upper (dorsal) part of some crustaceans and chelonians (tortoises and turtles). In crustaceans, it is made of chitin and is part of the **exoskeleton**. In chelonians, it consists of bony plates which are fused together. [2.3.a] [2.3.b]

■ **carbohydrate** Compound of carbon, hydrogen and oxygen in which the ratio of hydrogen to oxygen is 2:1 (the same as in water). **Cellulose, starch, glycogen** and all **sugars** are common carbohydrates. Digestible carbohydrates in the diet are a good source of energy. [3.7.b]

■ **carbon cycle** Passage of **carbon** from the air (as **carbon dioxide**) to plants by **photosynthesis** (forming **sugars** and **starches**), then through the **metabolism** of animals, to decomposition products which ultimately return to the atmosphere in the form of carbon dioxide. [2.7.a.] [3.7.a] [3.8.a]

■ **carbon dioxide** CO_2 Colourless gas formed by the combustion of carbon and its organic compounds, by the action of acids on **carbonates**, and as a product of **fermentation** and **respiration**. It is a raw material of **photosynthesis**. The accumulation of carbon dioxide in the atmosphere creates the

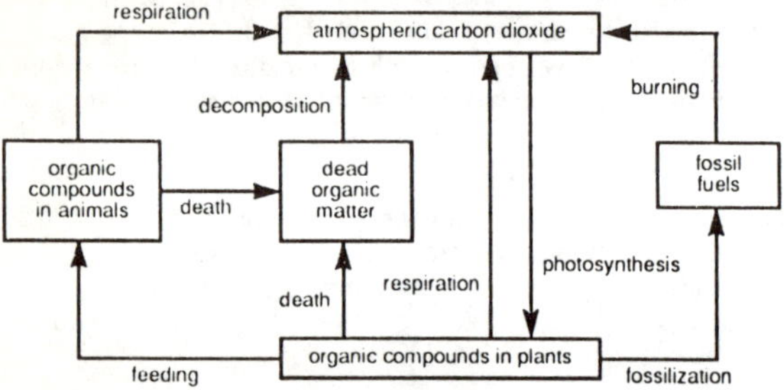

The carbon cycle

greenhouse effect. *See also* **carbon cycle**. [2.7.a] [3.7.a] [3.8.a]

■ **carcinogen** Substance capable of inducing a **cancer**. Most carcinogens are also **mutagens**, and this is thought to be their principal mode of action. [4.8.c]

■ **carcinoma** **Cancer** in epithelial tissue.

■ **cardiac** Relating to the **heart**. *See also* **cardiac muscle**. [3.4.a] [3.5.e] 3.8.a]

■ **cardiac muscle** Specialized **striated muscle**, found only in the heart, that continually contracts rhythmically and automatically without nervous stimulation (*i.e.* it is myogenic). [3.4.a] [3.5.e] [3.8.a]

■ **cardiogram** Alternative name for an **electrocardiogram**. [3.4.a] [3.5.e] [3.8.a]

■ **cardiovascular system Heart** and the network of blood vessels that circulate **blood** around the body. [3.4.a] [3.5.e] [3.8.a]

■ **carnassial tooth** Large **pre-molar** or **molar** tooth, found in **carnivores**, that is adapted for slicing food. [3.4.b]

■ **Carnivora** Order of meat-eating mammals, characterized by the possession of **carnassial teeth** and large **canine teeth**, which includes badgers, bears, cats, dogs, weasels and wolves. [2.3.b] [2.4.a] [2.5.b]

■ **carnivore** Meat-eating animal or a member of the **Carnivora**. *See also* **herbivore**; **omnivore**.

■ **carotene** Pigment that belongs to the **carotenoids**, a precursor of **retinol** (vitamin A).

■ **carotenoid** Member of a group of pigments found in *e.g.* carrots, which (like chlorophyll) can absorb light during **photosynthesis**. *See also* **carotene**.

■ **carotid artery** In vertebrates, either of two main **arteries** that carry blood from the heart to the head.

carotid body One of a pair of receptors located between the internal and external **carotid arteries**. It is sensitive to changes in the **carbon dioxide** and **oxygen** content of the blood, and sends impulses to the respiratory and blood-vascular centres in the brain to adjust these levels if necessary.

■ **carpal** Bone in the foot and wrist of **tetrapods**. There are 10–12 carpals in most animals; in human beings there are eight in each wrist.

■ **carpel** Female reproductive organ of a flowering plant, consisting of the **stigma**, **style** and **ovary**. Alternative name: gynaecium. [3.4.a] [3.4.d] [3.4.e] [3.8.a]

carpus The wrist, consisting of eight bones in human beings.

■ **carrier** *1*. Somebody who carries **pathogens** that cause disease, or carries genes that cause inherited disorders (*e.g.* haemophilia, colour blindness) and can pass them on to others, although not necessarily themselves having symptoms of the disorder. *2*. **Vector** (in medicine). *See also* **coenzyme**; **sex linkage**. [4.6.b] [4.8.b]

■ **cartilage** In animals that have a bony **skeleton**, pre-bone tissue occurring in young animals that becomes hard through **calcification**. It also occurs in some parts of the skeleton of adults, where it is structural (*e.g.* forming ear flaps and the larynx) or provides a cushioning effect during movement because it has a smoother surface and is softer than bone (*e.g.* intervertebral discs, cartilage at the ends of bones in joints). In some animals, *e.g.* sharks, rays and related fish (Chondrichthyes), the whole skeleton is composed of cartilage, even in the adult. [3.3.a] [3.5.e]

■ **cartilaginous fish** Fish that belongs to the class **Elasmobranchii** (Chondrichthyes); *e.g.* **sharks** and dogfish.

■ **casein Protein** that occurs in milk which serves to provide **amino acids** as nutrients for the young of mammals. It is the principal constituent of cheese.

■ **catabolism** Part of **metabolism** concerned with the breakdown of complex organic compounds into simple molecules for the release of energy in living organisms. *See also* **anabolism**.

■ **catalysis** Action of a **catalyst**.

■ **catalyst** Substance that alters the rate of a chemical reaction without itself undergoing any permanent chemical change. Sometimes, especially in biochemical reactions, a catalyst is essential to increase the rate sufficiently for the reaction to be detectable. *See also* **enzyme**.

catecholamine Member of a group of **amines** that include **neurotransmitters** (*e.g.* dopamine) and **hormones** (*e.g.* adrenalin).

CAT scan Abbreviation of computerized axial tomography scan, a technique for producing X-ray pictures of cross-sectional 'slices' through the brain and other parts of the body.

cauda equina Sheaf of nerve roots that arise from the lower end of the spinal cord and serve the lower parts of the body.

■ **caudal** Relating to the tail.

■ **caudal vertebra** One of the set of **vertebrae** nearest to the base of the **spinal column**.

■ **cell** Fundamental unit of living organisms. It consists of a membrane-bound compartment, often microscopic in size, containing jelly-like protoplasm. In plants each cell is bounded by a wall composed of **cellulose**, with the cell membrane pressed closely to it because of the **turgor** of the cell. In animals each cell is bounded by a cell membrane only. In all cells the **nucleus** contains **nucleic acids** (*e.g.* DNA) essential for the synthesis of new **proteins**. Cells may be highly specialized for their particular function and grouped into **tissues**, *e.g.* muscle cells. [3.5.a]

■ **cell body** Part of a nerve cell that contains the cell **nucleus** and other cell components from which an axon extends.

■ **cell differentiation** Way in which previously undifferentiated **cells** change structurally and take on specialized roles during growth and development (*e.g.* becoming liver cells or bone cells).

■ **cell division** Splitting of cells in two (**mitosis**) by division of the **nucleus** and division of the **cytoplasm** after duplication of the cell contents. *See also* **meiosis**. [4.5.a]

■ **cell membrane** *See* **plasma membrane**.

■ **cellulose** Major **polysaccharide** (*i.e.* a carbohydrate) of plants found in cell walls and in some **algae** and **fungi**. It is composed of **glucose** units aligned in long parallel chains, and gives cell walls their strength and rigidity. [2.4.a] [3.5.a]

■ **cell wall** Wall that surrounds plant cells, consisting mainly of **cellulose**. A membrane encloses the cell's **protoplasm**, and the wall is secreted through the membrane. It defines how large the cell can get, gives the cell support and rigidity, and is important for regulating the water content and fabric of a plant's vascular system. *See also* **lignin**. [2.4.a] [3.5.a]

■ **Celsius scale** Temperature scale on which the **freezing point** of pure water is 0°C and the **boiling point** is 100°C. It is the same as the formerly used centigrade scale. It was named after the Swedish astronomer Anders Celsius (1701–44).

■ **centigrade scale** Former name for the **Celsius scale**.

■ **central nervous system** (CNS) Concentration of nervous tissue responsible for co-ordination of the body. In **vertebrates** it is highly developed to form the **brain** and **spinal cord**. The CNS processes information from the sense organs and effects a response, *e.g.* muscle movement.

centrifuge Apparatus used for the separation of substances by **sedimentation** through rotation at high speeds, *e.g.* the separation of components of **cells**. The rate of sedimentation varies according to the size of the component. [3.7.c]

■ **centriole** Cylindrical body present in most animal **cells**. During **mitosis** it forms the poles of the **spindle**.

■ **centromere** Region on a **chromosome** that attaches to the **spindle** during **cell division**.

Cephalochordata Subphylum of the **Chordata** that consists of organisms with no vertebral column (*e.g.* amphioxus), although they do have a **notochord**.

cephalochordate Animal of the subphylum **Cephalochordata**.

cephalopod Member of the class **Cephalopoda**.

Cephalopoda Class of marine **molluscs** that includes cuttlefish, nautiloids, octopuses and squids (and extinct ammonites). Cephalopods have tentacles, highly developed eyes and nervous systems, and can move by forcing sea-water at high speed through their siphons.

cephalothorax Fused head and thorax (*e.g.* as in **arachnids**). Alternative name: prosoma.

■ **cerebellum** Front part of hindbrain of **vertebrates**. It is important in balance and muscular co-ordination. [3.7.c]

cerebral cortex Outer region of the **cerebral hemispheres** of the **brain** that contains densely packed **nerve cells** which are interconnected in a complex manner.

■ **cerebral hemisphere** In **vertebrates**, paired expansions of the anterior end of the forebrain. In mammals the forebrain is enormously enlarged by these expansions. [3.7.c]

cerebrospinal fluid (CSF) Liquid that fills the cavities of the **brain**, which are continuous with each other and the central canal of the **spinal cord**. The fluid nourishes the tissues of the **central nervous system** (CNS) and removes its secretions.

■ **cerebrum** Part of forebrain which in higher **vertebrates** expands to form the **cerebral hemispheres**.

cerumen Wax that forms in the **ear**.

■ **cervical** Relating to the neck or to the **cervix** of the womb. [3.4.c]

■ **cervical vertebra** **Vertebra** in the neck region of the **spinal column**, concerned with movements of the head.

■ **cervix** Neck, usually referring to the neck of the uterus (womb) at the inner end of the vagina. [3.4.c]

■ **CFC** Abbreviation of **chlorofluorocarbon**. [5.9.a]

Chaetognatha Phylum of invertebrate aquatic animals; the arrow-worms.

■ **character** In genetics, variation caused by a gene; an inherited trait. Alternative name: characteristic.

characteristic Typical feature (*e.g.* as in characteristics of living organisms). Alternative name: character.

■ **charcoal** Form of carbon made from incomplete burning of animal or vegetable matter.

■ **chemoreceptor** Cell that fires a nerve impulse in response to stimulation by a specific type of chemical substance; *e.g.* the taste buds on the tongue and olfactory bulbs in the nose contain chemoreceptors that provide the senses of taste and smell.

■ **chemotaxis** Response of organisms to chemical stimuli, *e.g.* the movement of **protozoa** towards nutrients.

■ **chemotherapy** Treatment of a disorder with drugs that are designed to destroy **pathogens** or cancerous tissue.

■ **chemotropism** Response of plants to chemical stimuli, *e.g.* the growth of the **pollen tube** towards the **ovary** in flowering plants.

■ **chiasma** *1.* Point along the **chromatid** of a homologous **chromosome** at which connections occur during crossing-over or exchange of genetic material in **meiosis**. *2.* Crossing-over point of the **optic nerves** in the brain.

chimaera *1.* Genetic mosaic or organism composed of genetically different **tissues** arising *e.g.* from **mutation** or mixing of cell types of different organisms. It can be achieved by incorporating donor cells at an early stage of embryonic development of the recipient. *2.* Member of a group of deep-sea cartilaginous fish. Alternative name: chimera.

■ **chitin Polysaccharide** (*i.e.* a carbohydrate) found in the exoskeleton of **arthropods**, giving it hard waxy properties.

■ **chlorination** Treatment of a substance with **chlorine** to bleach or disinfect it (*e.g.* drinking water).

■ **chlorofluorocarbon** (CFC) Fluorocarbon that has chlorine atoms in place of some of the fluorine atoms. Chlorofluorocarbons and fluorocarbons have similar properties, and are used as aerosol propellants and refrigerants (although blamed for damage to the **ozone layer**). [5.9.a]

■ **chlorophyll** Green pigment found in photosynthetic cells (contained in **chloroplasts**) of green plants. It is the major light-absorbing pigment and the site of the first stage of **photosynthesis**. *See also* **carotenoid**. [3.6.b] [3.7.a] [3.9.a]

Chlorophyta Largest division of **algae**, consisting of the green algae (which possess chlorophyll). Unicellular or multicellular, freshwater and marine-living, they exhibit **sexual** and **asexual reproduction**.

■ **chloroplast** Organelle found in photosynthetic cells of plants, containing the green pigment **chlorophyll**. Subsequent production of **carbohydrates** from **photosynthesis** occurs in chloroplasts. [3.6.b] [3.7.a] [3.9.a]

■ **cholesterol Sterol** found in animal tissues, a **lipid**-like substance that occurs normally in blood plasma, cell membranes and nerves, and which may accumulate to form gallstones. Abnormally high levels of cholesterol in the blood are connected with the onset of atherosclerosis, in which fatty materials are deposited in patches on artery walls and can restrict blood flow. Many **steroids** are derived from cholesterol.

choline $HOC_2H_4N(CH_3)_3OH$ Organic compound that is a constituent of the neurotransmitter **acetylcholine** and some **fats**. It is one of the B vitamins.

Chondrichthyes Alternative name for the fish class **Elasmobranchii**.

chondriosome Alternative name for **mitochondrion**.

■ **Chordata** Phylum of the animal kingdom that contains organisms with a **notochord, gill** slits and a hollow dorsal **nerve cord** at some stage of their development. Vertebrates are chordates, and thus the phylum includes mammals. [2.3.b] [2.5.b]

chordate Member of the phylum **Chordata**.

chorion *1.* Membrane that surrounds an implanted **blastocyst** and the embryo and foetus that develop from it in mammals, and the embryo in the eggs of reptiles and birds. *2.* Shell of an insect's egg.

chorionic gonadotrophin Hormone produced by the **placenta** during **pregnancy**. Its presence in the urine is the basis of many kinds of pregnancy testing.

chorionic villus sampling Testing during early pregnancy of small samples of tissue taken from the **chorion** for the presence of foetal abnormalities. *See also* **amniocentesis**.

■ **choroid** Layer of pigmented cells, rich in blood vessels, between the **retina** and **sclerotic** of the **eye**. [15.6.b]

■ **chromatid** One of two thread-like parts of a **chromosome**, visible during **prophase** of **meiosis** or **mitosis**, when the chromosome has duplicated. Chromatids are separated during **anaphase**. [4.7.a]

chromatin Basic **protein** that is associated with **eukaryotic chromosomes**, visible during certain stages of cell duplication.

■ **chromatography** Method of separating a mixture by carrying it in solution or in a gas stream through an absorbent material such as chromatography paper. The technique is an extremely important one in chemical and biological analysis.

chromatophore *1.* **Chromoplast**. *2.* In some vertebrates, a cell with pigment granules, which on movement of the granules alters the animal's colour. *3.* Structure in **prokaryotes** in which the **photosynthetic** pigments are located.

■ **chromoplast Organelle** in a plant cell that contains pigments, *e.g.* chromoplasts in carrots contain **carotenoids**.

■ **chromosome** Structure within the **nucleus** of a cell that contains **protein** and the genetic **DNA**. Chromosomes occur in pairs in **diploid** cells (ordinary body cells); **haploid** cells (gametes or sex cells) have only one of each pair of chromosomes in their nuclei. The number of chromosomes varies in different **species**. During mitotic cell division, each chromosome doubles and the two duplicates separate into the two new daughter cells. [4.7.a]

■ **chromosome map** Diagram showing the positions of various **genes** on appropriate **chromosomes**. [4.7.a] [4.9.d]

■ **chronic** Describing a condition or disorder that is long-standing (and often difficult to treat). *See also* **acute**.

■ **chrysalis** Alternative name for **pupa**.

chyle Milky fluid resulting from the absorption of fats in the **lacteals** of the small intestine; it is removed by the **lymphatic system**.

■ **chyme** Partly digested food that passes from the stomach into the duodenum and small intestine. [3.5.c] [3.7.b]

■ **ciliary body** Ring of tissue that surrounds the lens of the eye. It generates the **aqueous humour** and contains ciliary muscles, which are used in **accommodation**. [15.6.b]

■ **ciliary muscle** Muscle that controls the shape of the lens of the **eye** and thus achieves **accommodation**. [15.6.b]

■ **cilium** (plural cilia) Small hair-like structure that moves rhythmically on the surface of a cell or the whole **epithelium**. Cilia usually cover the surface of a cell and cause movement in the fluid surrounding it. For example, ciliated cells line the respiratory tract, wafting mucus out of the bronchioles and bronchi. Cilia are also used for locomotion by some single-celled aquatic organisms. *See also* **flagellum**.

■ **circadian rhythm** Cyclical variation in the physiological, metabolic or behavioural aspects of an organism over a period of about 24 hours; *e.g.* sleep patterns. It may arise from inside an organism or be a response to a regular cycle of some external variation in the environment. Alternative term: diurnal rhythm. *See also* **biological clock**.

■ **circulatory system** In animals, transport system that maintains a constant flow of **tissue fluid** in sealed vessels to all parts of body. *E.g.* in the **blood vascular system**, oxygen and food materials dissolved in blood diffuse into each cell; waste products, including carbon dioxide, diffuse out of the cells and into the blood. *See also* **lymphatic system**. [3.4.a] [3.5.e] [3.8.a]

cistron Functional unit of a **DNA** chain that controls **protein** manufacture.

■ **citric acid** $C_6H_8O_7$ Hydroxy-tri**carboxylic acid**, present in the juices of fruits and made by fermenting residues from sugar refining. It is important in the **Krebs cycle**, and is much used as a flavouring and in medicines.

■ **citric acid cycle** Alternative name for **Krebs cycle**. [3.9.a]

■ **class** In biological **classification**, one of the groups into which a phylum is divided, and which is itself divided into orders; *e.g.* Mammalia (mammals), Aves (birds). [2.3.b] [2.4.a] [2.5.b]

■ **classification** In biology, the placing of living organisms into a series of groups according to similarities in structure, physiology, genotype, and other characteristics. The smallest group is the **species**. Similar species are placed in a genus, similar genera are grouped into families, families into orders, orders into classes, classes into phyla (or divisions in plants), and phyla into **kingdoms**. There may also be intermediate groups such as suborders, subclasses and subphyla. Modern classification is usually intended to reflect degrees of evolutionary relationship, although not all experts agree on single classification schemes for animals or for plants. *See also* **binomial nomenclature**. [2.3.b] [2.4.a] [2.5.b]

■ **clavicle** In vertebrates, the anterior bone of the ventral side of the shoulder girdle. Alternative name: collarbone. [3.3.a]

cleavage *1.* In biology, a series of **mitotic** divisions of a fertilized **ovum**. *2.* In biochemistry, the splitting of chemical bonds (*e.g.* protease enzymes cleave peptide bonds from proteins to release amino acid residues).

■ **climax community** Complex but stable plant community that can perpetually regenerate itself under prevailing environmental conditions. It is the final stage of plant **succession**. Alternative name: climax vegetation.

■ **clinical thermometer** Type of (mercury) **thermometer** used for taking body temperature. [3.8.a]

■ **clinostat** Horizontal or vertical turntable used to rotate plants so as to counteract the effect of a stimulus (*e.g.* gravity) that normally acts in only one direction.

■ **clitoris** Mass of erectile tissue in female mammals, the equivalent of the **penis** in males, situated in front of the opening of the **vagina**. [3.4.c]

cloaca *1.* In most non-mammalian vertebrates, the single posterior opening to the body. Into it open the anus, reproductive ducts and urinary ducts. *2.* Terminal part of the intestine in some invertebrates, *e.g.* sea cucumbers.

■ **clone** One of many descendants produced by **vegetative propagation** from one original plant seedling, or **asexual reproduction** or **parthenogenesis** from a single animal. Members of a clone have identical genetic constitution. [4.9.c]

■ **clotting factor** **Protein** structure, *e.g.* thrombin and fibrinogen, that induces blood **coagulation** when a blood vessel is broken.

Cnidaria Phylum of invertebrate animals, which includes corals, hydra, jellyfish and sea-anemones. They were formerly classified as **Coelenterata**.

cnidarian Animal that is a member of the phylum **Cnidaria**.

■ **CNS** Abbreviation of **central nervous system**. [3.5.c] [3.7.c]

■ **coagulation** *1.* Process by which bleeding is arrested. Thrombin is produced in the absence of antithrombin from the combination of prothrombin and calcium ions. This interacts with soluble fibrinogen to precipitate it as the insoluble blood protein fibrin, which forms a mesh of fine threads over the wound. Blood cells become trapped in the mesh and form a clot. *2.* Irreversible setting of protoplasm on exposure to heat or poison. *3.* Precipitation of colloids, *e.g.* proteins, from solutions. [3.5.e] [3.8.b] [4.10.a]

■ **coccus** Spherical-shaped **bacterium**. Cocci may join together to form clumps (staphylococci) or chains (streptococci). [3.5.e]

■ **coccyx** Bony structure in primates and amphibians, formed by fusion of tail vertebrae; *e.g.* in human beings it consists of three to five vestigial vertebrae at the base of the spine. [3.5.e]

■ **cochlea** Spirally coiled part of the inner **ear** in mammals. It translates sound-induced vibrations into nerve impulses that travel along the auditory nerve to the brain, where they are interpreted as sounds. [3.4.a] [3.5.e] [3.8.a]

■ **codeine** Pain-killing drug (a narcotic analgesic), the methyl derivative of **morphine**. [3.4.b]

co-dominant Describing a gene condition in which neither **allele** is **dominant** or **recessive** (*e.g.* A, B and AB blood groups).

■ **Coelenterata** Major group of invertebrate animals, formerly given the status of a phylum but now divided into Cnidaria and Ctenophora. Coelenterates have a body made up of two layers of cells, **ectoderm** and **endoderm**, which are separated by a middle layer of jelly, the **mesoglea**. They are symmetrical, and have a single body cavity with one opening at the mouth. The nervous system is a diffuse network, and they have neither excretory nor blood systems. Common coelenterates include jellyfish, corals and hydra. [2.5.b]

coelenterate Member of the **Coelenterata**.

■ **coelom** *1.* Body cavity of **triploblastic** animals (*e.g.* vertebrates), formed from the **mesoderm**. It contains coelomic fluid, in which colourless **phagocytic** corpuscles as well as the gut are suspended. Coelomic fluid keeps the body moist for respiration and the corpuscles keep the space free from **bacteria**. *2.* Body cavity of an insect.

coenocyte Multinucleate mass of **protoplasm** formed by the division of the **nucleus** from a cell that has only a single nucleus, *e.g.* as in **fungi** and **algae**.

■ **coenzyme** Organic compound essential to catalytic activities of **enzymes** without being utilized in the reaction. Coenzymes usually act as carriers of intermediate products, *e.g.* ATP.

■ **cold-blooded** Alternative name for **poikilothermic**.

Coleoptera Beetles, a major order of **insects**. *See* **beetle**.

collagen Fibrous protein connective tissue that binds together bones, ligaments, cartilage, muscles and skin.

■ **collarbone** Alternative name for the **clavicle**. [3.3.a] [3.4.a]

■ **colloid** Form of matter that consists of extremely small particles, about 10^{-4} to 10^{-6} mm across, so small that they remain suspended and dispersed in a medium such as air or water. Common colloids include aerosols (*e.g.* fog, mist), emulsions (*e.g.* milk) and gels (*e.g.* gelatin, rubber). Mucus may also be colloidal, and clay consists of mainly colloidal-sized soil particles. A non-colloidal substance is termed a crystalloid. [5.7.a] [7.8.b]

■ **colon** *1.* Large intestine of vertebrates, in which the main function is the absorption of water from **faeces**. *2.* In insects, the wide posterior part of the hind gut. [3.4.a] [3.5.e] [3.8.a]

colonization Process by which species begin to live in new **habitats**, usually through natural means or sometimes through introduction by human activity.

■ **colony** *1.* Bacterial growth on a solid medium that forms a visible mass. *2.* Growth of a group of individual plants of one species that invade new ground. *3.* Collection of individuals that live together with some degree of interdependence (*e.g.* social insects such as some ants, bees and wasps).

■ **colostrum** Yellowish milky fluid secreted from the mammalian breast immediately before and after childbirth. It contains more **antibodies** and **leucocytes** (and less fat and carbohydrates) than true milk, which follows within a few days. [3.6.c] [3.9.b]

■ **colour** Visual sensation or perception that results from the adsorption of light energy of a particular **wavelength** by the **cones** of the **retina** of the eye. There are two or more types of cone, each of which is sensitive to different wavelengths of light. The brain combines nerve impulses from these cones to produce the perception of colour. The colour of an object thus depends on the wavelength of light it reflects (other wavelengths being absorbed) or transmits.

■ **colour blindness** Inability to distinguish between certain colours. It is a **congenital** abnormality that affects 6% of human males and 1% of females. The most common defect is red-green colour blindness, which results in observation of both colours in grey, blue or yellow, depending on the amount of yellow and blue present in the light. [4.6.b] [4.8.b] [15.6.c]

■ **commensalism** Close relationship between two plant or animal organisms or communities from which one usually benefits more than the other (which is nevertheless unharmed by the association); *e.g.* a sea anemone living on the shell of a hermit crab is carried around and benefits from bits of food that float away from the crab while it is feeding. *See also* **parasitism**; **symbiosis**.

■ **community** In biology, collection of interacting but different species that occupy the same **habitat**.

compact Describing a condition that results from physical pressure (*e.g.* a compact soil holds less water and air than a non-compact one).

■ **competition** In biology, the struggle within a **community** between organisms of the same or different species for survival. *See also* **natural selection**. [2.5.d] [2.6.a]

compost Gardening term for a collection of leaves, weeds and other organic debris that is left to decay and then used as fertilizer. [2.8.a]

■ **compound eye** One of the paired eyes of most adult arthropods (*e.g.* insects), consisting of many units (called ommatidia) each with its own lens. A fly, for example, can detect movement very easily as an image moves from one ommatidium to the next.

■ **conception** Fertilization of a human egg, usually taken to mean also **implantation** in the womb of the resultant **zygote** (which goes on to develop into an embryo and foetus). [3.6.f]

■ **conditioned reflex** Animal's response to a neutral stimulus that learning has associated with a particular effect; *e.g.* a laboratory rat may learn (be conditioned) to press a lever when hungry because it associates this action with receiving food. [3.5.e] [3.7.c]

■ **cone** *1.* Light-sensitive nerve cell present in the **retina** of the eye of most vertebrates; it can detect colour. *See also* **rod.** *2.* Reproductive structure of **gymnosperms** (*e.g.* pines and other conifers). [15.6.b]

■ **congenital** Dating from birth or before birth. Congenital conditions may be caused by enviromental factors or be inherited. [3.4.c] [3.6.c]

■ **conifer** Plant that is a member of the order **Coniferales.**

■ **Coniferales** Large order of cone-bearing, usually evergreen shrubs or trees; conifers. They include pines and firs. Larch is the only European conifer that is deciduous. [2.3.a] [2.5.a]

conjugation *1.* In biology, form of reproduction that involves the permanent or temporary union of two isogametes, *e.g.* in certain green **algae**. In **protozoa**, two individuals partly fuse, exchanging nuclear materials. When separated, each cell divides further to give new individuals or, more usually, uninucleate **spores** (which eventually develop into full adults).

connective tissue Strong **tissue** that binds **organs** and tissues together. It consists of a **glycoprotein** matrix containing **collagen** in which cells, fibres and vessels are embedded. The most widespread connective tissue is **areolar tissue**.

■ **contraception** Avoidance of **conception**, by means of a method or device that is designed to prevent male sperm from reaching the female egg (ovum) during or soon after intercourse, or by means of **hormones** (the contraceptive pill) that suppress **ovulation** by interfering with the woman's **menstrual cycle**. [3.6.f]

■ **contractile vacuole** Membranous sac within a single-celled organism (*e.g.* amoebae and other protozoans) which fills with water and suddenly contracts, expelling its contents from the cell. It carries out **osmoregulation** and excretion.

control Part of a scientific experiment in which the experimental conditions are checked against standard (*i.e.* non-experimental) ones.

convergent evolution Tendency of species that live in a single uniform environment to develop similar characteristics.

Copepoda Subclass of **Crustacea**, many of which are minute parasitic marine animals. They also include the non-parasitic, freshwater *Daphnia*, commonly known as the water flea.

copepod Crustacean that is a member of the subclass **Copepoda**.

■ **coral** Substance containing calcium carbonate secreted by various marine organisms (*e.g.* Anthozoa) for support and habitation.

■ **cork** *1.* Layer of dead cells on the outside of plant stems and roots. It is impermeable to air and water, and so protects the living cells inside against water loss and physical injury. In the cork oak this layer is exceptionally thick and can be stripped off, to be used for making bottle stoppers ('corks') or heat-resistant insulating material. Alternative name: phellem. *See also* **lenticel**. [3.4.a]

cork cambium *See* **phellogen**.

■ **corm** Underground organ developed by some plants. It is formed from **stem** tissue, and has adventitious roots growing from its lower surface. Primarily a storage organ, it may also play a part in **vegetative propagation** by producing buds that develop into new plants. [3.4.a] [3.5.e] [3.8.a]

■ **cornea** Transparent connective tissue at the front surface of the **eye** of vertebrates, overlying the **iris**. Together with the **lens**, it focuses incoming light onto the **retina**. [3.4.a] [3.5.e] [3.8.a]

cornification Alternative name for **keratinization**.

■ **corolla** Inner part of a flower, consisting of petals. It is often coloured. [3.4.a] [3.5.e] [3.8.a]

■ **coronary vessels** **Arteries** and **veins** that carry the blood supplying the heart muscle in vertebrates.

corpus callosum Thick bundle of nerve fibres in the middle of the **brain** that connects the two cerebral hemispheres.

■ **corpus luteum** Yellow body formed in the **ovary** of female mammals which produces the hormone **progesterone**. It develops from the **Graafian follicle** after **ovulation**. If **fertilization** does not occur, the corpus luteum degenerates. [3.4.c] [3.5.e]

■ **cortex** Outer layer of a structure, *e.g.* the rind of plant parenchyma tissue surrounding the vascular cylinder in the stem and roots of plants, or the outer layer of cells of the **adrenal gland** or **brain** of mammals. *See also* **medulla**.

corticosteroid **Hormone** secreted by the **adrenal cortex**, which controls sodium and water **metabolism** as well as **glycogen** formation.

corticotrophin Alternative name for **adrenocorticotrophic hormone** (ACTH).

■ **cortisone Hormone** isolated from the **adrenal cortex**, used in the treatment of severe hay fever, rheumatoid arthritis and other inflammatory conditions. [3.8.a]

costal Concerning the ribs.

■ **cotyledon** *1*. Leaf that forms part of the embryo of seeds, usually lacking chlorophyll (until exposed to light after germination). Monocotyledons have one and dicotyledons have two cotyledons in each seed. In certain plants, *e.g.* peas and beans, these are food-storage organs. Cotyledons of many plants appear above ground, develop chlorophyll and synthesize food material by **photosynthesis**. *2*. One of the leaves of the embryo in flowering plants. *3*. Leaf developed by a young fern plant. [3.4.d]

■ **Cowper's gland** One of a pair of small glands located below the **prostate gland** and connected to the **urethra**; it produces fluid for **semen**. It was named after the British anatomist William Cowper (1666–1709). [3.4.c]

Coxsackie virus Member of a group of **viruses** that cause inflammatory diseases in human beings. It was named after the city in New York state where it was first found.

■ **cranial** Relating to the **cranium** and **brain**.

■ **cranial nerve** One of the 10 to 12 pairs of nerves (12 pairs in human beings) connected directly with a vertebrate's brain and supplying the sense organs, muscles of the head and neck and abdominal organs. Together with the spinal nerves they make up the peripheral nervous system. [3.7.c]

Craniata Alternative name for **Vertebrata**.

■ **cranium** Part of the **skull** that encloses and protects the brain, consisting of eight fused bones in human beings.

creatine $NH_2C(NH)N(CH_3)CH_2COOH$ White crystalline amino acid present in muscle, where it plays an important role in muscle contraction. It is broken down to **creatinine**.

creatinine $C_4H_7N_3O$ Heterocyclic crystalline solid formed by the breakdown of **creatine** and excreted in urine.

Cro-Magnon man Earliest recognized form of modern man. Cro-Magnons are known to have lived in Europe 30–40,000 years ago, but may have appeared in Africa or Asia at an earlier date.

■ **crossing over** Exchange of material between homologous **chromosomes** during **meiosis** (cell division). It is the mechanism that alters the pattern of genes in the chromosomes of offspring, giving the genetic variation associated with **sexual reproduction**. [4.7.a]

■ **Crustacea** Large class of **arthropods** which includes barnacles, crabs, lobsters, prawns and woodlice. Characterized by having two pairs of **antennae**, most crustaceans are aquatic. [2.3.b] [2.4.a] [2.5.b]

crustacean Member of the class **Crustacea**.

crystalloid Substance that is not a **colloid** and can therefore pass through a semipermeable membrane.

CS gas $C_6H_4ClCH=C(CN)_2$ White irritant organic compound used as a tear gas for riot control. Alternative name: (2-chlorobenzylidene)-malononitrile.

■ **cusp** In anatomy, pointed part of a tooth. [3.7.b]

■ **cuticle** *1.* In plants, deposit of waxy waterproof material that forms a continuous layer over aerial parts, broken only by **stomata**. It provides protection against water loss and injury. *2.* In animals, protective layer of hard material that covers many arthropods (*e.g.* crustaceans, insects), where it may form the **exoskeleton**. [3.4.a] [3.5.e] [3.7.a]

cutin Mixture resulting from oxidation and condensation of fatty acids which is deposited on or in the outer layer of cell walls (cuticle) of plants.

Cyanobacteria Alternative name for the **Cyanophyta**.

Cyanophyta Division of single-celled, photosynthetic, **prokaryotic** organisms. Cyanophytes sometimes join together in colonies, and are most commonly found in water. Alternative names: blue-green algae; Cyanobacteria; Cyanophyceae.

cyanophyte Member of the **Cyanophyta**.

■ **cybernetics** Science that employs control systems resembling those of living things for mechanisms and electronic systems (*e.g.* in building industrial robots).

cystine $(SCH_2CH(NH_2)COOH)_2$ Dimeric form of the **amino acid** cysteine, found in **keratin**.

cytochrome Respiratory pigment found in organisms that use **aerobic respiration**.

cytokinin Plant **hormone** that stimulates the division of plant cells.

■ **cytology** Study of **cells**. [3.5.a]

■ **cytoplasm** **Protoplasm** of a cell other than that of the **nucleus**. [3.5.a]

cytosine Colourless crystalline compound, derived from **pyrimidine**. It is a major constituent of **DNA** and **RNA**.

cytotoxic Describing a drug that destroys or prevents the replication of cells, used in **chemotherapy** to treat **cancer**.

D

■ **Darwinism** Theory of **evolution** which states that living organisms arise in their different forms by gradual change over many generations, and that this process is governed by **natural selection**. It was proposed by the British naturalist Charles Darwin (1809–82). [4.9.a]

■ **daughter cell** One of the two cells produced when a plant or animal cell divides by **mitosis**.

■ **daughter nucleus** One of the two nuclei produced when the nucleus of a cell divides.

■ **DDT** Abbreviation of dichlorophenyltrichloroethane. This compound was once much used as an insecticide, but is now banned in most countries because of its high toxicity and the fact that it can pass through food chains without degrading. [2.7.b] [2.9.c] [2.10.b]

deaminase Enzyme that catalyses the removal of an **amino group** ($-NH_2$) from an organic molecule.

■ **deamination** Enzymatic removal of an **amino group** ($-NH_2$) from a compound. The process is important in the breakdown of **amino acids** in the liver. Ammonia formed by deamination is converted to **urea**, carried in the bloodstream to the kidneys, and excreted in **urine**. [3.7.b]

decapod Member of the crustacean order **Decapoda**.

Decapoda *1*. Large order of **crustaceans** that includes crabs, lobsters, crayfish and shrimps. There are more than 8,000 species of decapods, most of which are marine. *2*. Sub-order of **cephalopods** that includes squids and cuttlefish.

■ **decay** Natural breakdown of an organic substance; **decomposition**. [2.4.b] [5.2.a] [5.4.a] [5.5.b]

■ **deciduous** Shedding leaves at certain seasons of the year. The process occurs most commonly in autumn for plants in temperate regions. *See also* **evergreen**. [2.1.a] [2.3.a] [2.5.a]

■ **decomposer Saprophyte** organism that breaks down organic materials into simple molecules, *e.g.* **fungi**. [5.2.a] [5.4.a]

decomposition *1.* Rotting of a dead organism, often brought about by **bacteria** or **fungi**. *2.* Breaking down of a chemical compound into its component parts (elements or simpler substances), often brought about by heat or light. [3.5.d] [5.4.a] [5.5.b]

■ **deficiency disease** Disorder brought about by the lack of a certain food substance in the diet, *e.g.* a vitamin, mineral or amino acid. [3.7.b] [3.8.b] [3.10.a]

■ **deforestation** Loss of forest trees by felling or by the action of erosion or **acid rain**. Because it results in there being fewer trees to convert atmospheric carbon dioxide into oxygen (by **photosynthesis**), deforestation on a large scale contributes to the **greenhouse effect**. *See also* **afforestation**. [2.6.b]

deglutition Another term for swallowing.

dehiscene Opening to liberate seeds; in particular, the fruits of a plant, *e.g.* poppy.

■ **dehiscent** Term that describes seeds and berries that burst open when ripe or mature.

■ **dehydration** Loss or removal of water which, in a living organism, results in a reduction of tissue fluid, possibly to a harmful level. [3.8.a]

dehydrogenase Enzyme that catalyses the removal of hydrogen from a compound, and thus causes the compound's **oxidation**.

deme Any distinct population of interbreeding organisms that share a set of characteristics. A deme is a sub-division of a species that is much more localized but less differentiated than a **subspecies**. *See also* **race**; **variety**.

■ **denature** To unfold the structure of the **polypeptide** chain of a **protein** by exposing it to a higher temperature or extremes of **pH**. This results in the loss of biological activity and a decrease in solubility.

■ **dendrite** In biology, an elongated process from a **nerve cell** that links it with another nerve cell. Alternative name: dendron. [3.7.c]

dendrochronology Science of dating based on annual growth rings in trees. Variations in climate produce variations in ring width, and researchers have been able to establish patterns of these variations in certain regions that go back as much as 3,000 years.

■ **dendron** Thread-like outgrowth of a **nerve cell** (neurone) that carries impulses away from the cell body. Alternative name for **dendrite**. *See also* **axon**. [3.7.c]

■ **denitrification** Process that occurs in soil, in which bacteria break down **nitrates** and **nitrites**, with the liberation of **nitrogen**. [5.4.a] [5.7.a]

■ **dental caries** Decay of teeth caused by bacteria that live on food debris in plaque. The bacteria produce acid (as a waste product) that dissolves calcium salts in the tooth **enamel**. [3.4.b]

■ **dental formula** Notation that shows the number of each kind of tooth possessed by a **mammal**. The number of teeth in one side of the jaw is given, with the number in the upper jaw before that in the lower jaw, and in the order incisors, canines, premolars, molars; *e.g.* the human formula is 2/2 1/1 2/2 3/3. [3.4.b]

■ **dentine** Layer that occurs under the **enamel** of teeth. It is similar to bone, but harder, and is perforated by thin extensions from tooth-forming cells. [3.4.b]

■ **deoxyribonucleic acid** (DNA) Nucleic acid that is usually referred to by its abbreviation. *See* **DNA**. [4.9.d]

deoxyribonucleotides Compounds that are the fundamental units of **DNA** (deoxyribonucleic acid). Each **nucleotide** contains a nitrogenous **base**, a pentose **sugar** and **phosphoric acid**. The four bases characteristic of deoxyribonucleotides are **adenine, guanine, cytosine** and **thymine**.

Dermaptera Order of winged insects that consists of the earwigs. There are about 1,800 species world-wide.

■ **dermis** Layer of skin that is the innermost of the two main layers. It is composed of **connective tissue** and contains blood, lymph vessels, sensory nerves, hair follicles, sweat glands and some muscle cells. *See also* **epidermis**. [3.4.a] [3.4.b] [3.5.e]

Dermoptera Small sub-order of the **insectivores** that includes the 'flying lemur', the most highly developed of all gliding mammals.

desmid Member of the Desmidaceae family of unicellular green **algae**, usually found in unpolluted freshwater habitats and **plankton**.

desquamation Shedding of the skin, as when a snake or lizard sloughs its skin.

■ **detergent** Substance that is used as a cleaning agent. Detergents are particularly useful for cleaning because they lower **surface tension** and emulsify fats and oils, allowing them to go into solution with water without forming a scum with any of the substances that cause **hardness of water**. Soaps act in a similar way, but form an insoluble scum in hard water. Detergents may, however, be a source of **pollution** in rivers and lakes.

determinant *1.* In medicine and biology, region or regions of an **antigen** molecule required for its 'recognition' (binding) by a particle or **antibody**. The selective nature of this molecular interaction confers specificity on the immune reaction of the antibody-producer. *2.* Also in biology, a factor that transmits inherited characteristics, *e.g.* a **gene**.

dextrin Polysaccharide of intermediate chain length produced from the action of **amylases** on **starch**. It is used as a gum and thickening agent.

dextrose Alternative name for **glucose**.

■ **diabetes** Disorder caused by a lack of a hormone **(insulin)**, which is normally secreted by the islets of Langerhans in the pancreas to control the levels of sugar (glucose) in the blood. Full name: diabetes mellitus. [4.6.b] [4.8.b]

diakinesis Phase of **cell division** that occurs at the final stage of **prophase** of the first division in **meiosis**. During this phase the **chromosomes** become short and thick, forming more **chiasmata**, the **nucleoli** and nuclear membrane disappear, and the **spindle** appears for the process of division.

dialysed iron Colloidal solution of iron(III) hydroxide, $Fe(OH)_3$. It is a red liquid, used in medicine.

■ **dialysis** Separation of **colloids** from **crystalloids** using selective diffusion through a semipermeable membrane. It is the process by which globular **proteins** can be separated from low-molecular-weight solutes, as in filtering ('purifying') blood in an artificial kidney machine: the membrane retains protein molecules and allows small solute molecules and water to pass through. [3.8.b] [3.8.c]

diapause Pause in the development of an individual insect that may occur at any stage of growth – egg, larva or pupa. Diapause is usually induced by an adverse seasonal change, and the organism postpones development until conditions become more favourable.

■ **diaphragm** *1.* In anatomy, a sheet of **muscle** present in mammals, located below the **lungs**. It is attached to the body wall at the sides, and separates the **thorax** from the **abdomen**. During **respiration** the muscle contracts and relaxes, so forming an important part of the mechanism for filling and emptying the lungs. *2.* A birth-control device fitted over the entrance to the uterus to prevent the entry of sperm (alternative name: Dutch cap). [3.4.a] [3.5.e] [3.8.c]

diaspore Structure that functions as a means of dispersal for plant and fungus species, *e.g.* a **seed** or a **spore**.

diastase Alternative name for **amylase**.

■ **diastema** Gap in the jaw of mammals (usually **herbivores**) where there are no teeth. It permits manipulation of leafy food by the animal's tongue. [3.7.b]

■ **diastole** Phase of the **heartbeat** in which the heart undergoes relaxation and refills with blood from the veins. The term also applies to a contractile vacuole in a cell when it refills with fluid. *See also* **systole**.

diatom Alternative name for a member of the **Bacillariophyta**.

■ **dichotomous** Of plants, divided into two equal branches.

■ **dicotyledon** Flowering plant that has two seed leaves, broad-veined leaves and stems with **vascular bundles**. *See also* **monocotyledon**. [2.5.b]

diet Food that is eaten according to certain criteria. *See also* **balanced diet**. [3.4.b]

■ **diffusion** Process by which gases or liquids mix together – *e.g.* in gas exchange between plant leaves and air. *See also* **active transport**; **dialysis**; **osmosis**. [3.7.b]

■ **digestion** Breakdown of complex substances in food by **enzymes** in the **alimentary canal** to produce simpler soluble compounds, which pass into the body by **absorption** and **assimilation**. Ultimately **carbohydrates** (*e.g.* starch, sugar) are broken down to glucose, **proteins** to amino acids, and **fats** to fatty acids and glycerol. [3.4.a] [3.5.e] [3.8.a]

■ **digit** In biology, a finger or toe, or other analogous structure.

■ **digitalis** Potent **alkaloid** that is extracted from plants of the genus *Digitalis* (foxgloves). It is used in medicine as a heart stimulant. [3.8.b]

dilate To widen; to produce **dilation**.

■ **dilation** In biology and medicine, the widening or expansion of an organ, opening, passage or vessel (*e.g.* of the **cervix** during birth). Alternative term: dilatation. [3.6.c] [3.8.a]

■ **dimorphism** Existence of two distinct forms of an organism; *e.g.* sexual dimorphism in some animals, aerial and submerged leaves of some aquatic plants.

■ **dinosaur** Extinct reptile which existed during the **Mesozoic** era. Dinosaurs were a successful and diverse group that dominated the terrestrial environment of Earth for

140 million years. Climatic changes induced by continental shifts are thought by many to have caused their extinction about 65 million years ago at the end of the **Cretaceous** period, although other theories give different reasons. [2.4.c]

dioecious Describing plants in which the male and female reproductive organs are borne on different individuals or parts.

dioxan $(CH_2)_2O_2$ Colourless liquid cyclic **ether**. It is inert to many reagents and frequently used in mixtures with water to increase the solubility of organic compounds such as **alkyl halides**. Alternative names: 1,4-dioxan.

dioxin $C_{12}H_4Cl_4O_2$ Highly toxic by-product of organic synthesis, used as a defoliant and herbicide, which in even small doses can cause allergic skin reactions.

■ **diploblastic** Describing an animal in which the body wall consists of two cellular layers: the outer **ectoderm** and the inner **endoderm**, sometimes separated by a middle **mesoglea** layer. No organs develop from the latter, but only from the ectoderm and endoderm, as in *e.g.* coelenterates.

■ **diploid** In a cell or organism, describing the existence of **chromosomes** in homologous pairs, *i.e.* twice the **haploid** number ($2n$). It is characteristic of all animal cells except **gametes**. In lower plants exhibiting **alternation of generations**, the **sporophyte** is diploid and the **gametophyte** is haploid. [4.7.a]

diplotene In **meiosis**, the stage in late prophase when the pairs of **chromatids** begin to separate from the **tetrad** formed by the association of homologous **chromosomes**. **Chiasmata** can be seen at this stage.

directive body Alternative name for **polar body**.

■ **disaccharide** Sugar with molecules that consist of two **monosaccharide** units linked by **glycoside** bonds; *e.g.* **sucrose, maltose, lactose.**

■ **dispersal** Means by which a plant's seeds are scattered; *e.g.* by the wind, stuck to the fur of animals, eaten (with fruit) by birds and even by flowing water.

■ **diurnal** During the day, daily (*e.g.* a diurnal animal is active by day).

■ **division** *1.* In biological **classification,** one of the major groups into which the plant kingdom is divided. The members of the group, although often quite different in form and structure, share certain common features, e.g. bryophytes include the mosses and liverworts. Divisions are divided into classes, often with an intermediate subdivision. The equivalent of a division in the animal kingdom is a **phylum**. *2.* In biology, the formation of a pair of **daughter cells** from a parent cell (*see* **cell division**).

dl-form Term indicating that a mixture contains the **dextrorotatory** and the **laevorotatory** forms of an optically active compound in equal molecular proportions.

■ **DNA** Abbreviation of deoxyribonucleic acid, the long thread-like molecule that consists of a double helix of **polynucleotides** held together by **hydrogen bonds**. DNA is found chiefly in **chromosomes**, and is the material that carries the hereditary information of all living organisms (although most, but not all, viruses have only ribonucleic acid, **RNA**). [4.9.d]

DNA hybridization Technique in which **DNA** from one species is induced to undergo base pairing with DNA or **RNA** from another species to produce a hybrid DNA (a process known as annealing).

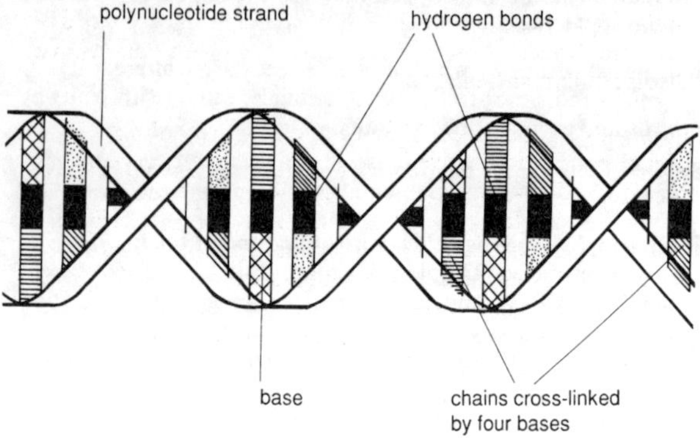

Structure of DNA

■ **dominant** In a **heterozygous** organism, describing the **gene** that prevents the expression of a **recessive** allele in a pair of homologous **chromosomes**. Thus the **phenotype** of an organism with a combination of dominant and recessive genes is similar to that with two dominant alleles. [4.8.a] [4.8.c] [4.10.a]

■ **donor** In medicine, person or animal that donates blood, tissue or organs for use by another person or animal.

dopa Amino acid derivative that is a precursor in the synthesis of **dopamine** and is laevorotatory (L-dopa). Found particularly in the adrenal gland and in some types of beans, it is used in the treatment of Parkinson's disease. Alternative name: dihydroxyphenylalanine.

dopamine Precursor in the synthesis of **adrenalin** and **noradrenalin** in animals. It is found in highest concentration in the corpus striatum of the brain, where it functions as a neurotransmitter. Low levels are associated with Parkinson's disease in human beings.

■ **dormancy** Period of minimal metabolic activity of an organism or reproductive body. It is a means of surviving a period of adverse environmental conditions, *e.g.* cold or drought. Examples of some dormant structures are spores, cysts and perennating organs of plants. Environmental factors such as day length and temperature control both the onset and ending of dormancy. Dormancy may also be prompted and terminated by the action of hormones, *e.g.* abscinic acid and **gibberellins**, respectively. *See also* **aestivation**; **hibernation**. [2.3.c]

■ **dorsal** Describing the upper surface of an organism. In vertebrates this is the surface nearest to the backbone. In plants the dorsal surface of a leaf is the upper surface, usually with the thicker **cuticle**.

dosimeter Instrument that measures the dose of **radiation** received by an individual or an area.

■ **double recessive** Homozygous condition in which two **recessive alleles** of a particular **gene** are at the same locus on a pair of **homologous chromosomes**, so that the recessive form of the gene is expressed in the **phenotype**. [4.8.a]

■ **Down's syndrome** Abnormal chromosomal condition caused by the presence of an extra **autosomal** chromosome 21. It is characterized by abnormal physical development and mental retardation. Alternative names: mongolism, trisomy 21. [4.8.b] [4.8.c]

■ **drug** Chemical (often a biochemical) that has a stimulating, narcotic or healing effect on the body. *See also* **addiction**; **analgesic**; **antibiotic**. [3.4.b]

■ **drupe** Fleshy fruit covered by **epicarp** and containing one or more seeds surrounded by a hard stony wall, the **endocarp**. Drupes with one seed include plums and cherries; many-seeded drupes include holly and elder fruits. Blackberries and raspberries are collections of small drupes or drupelets. Alternative name: pyrenocarp. *See also* **berry**.

■ **ductless gland** Alternative name for **endocrine gland**. [3.8.a]

■ **duodenum** First section of the **small intestine** which is mainly secretory in function, producing digestive **enzymes**. It also receives pancreatic juice from the **pancreas** and bile from the gall bladder. [3.4.a] [3.5.e] [3.8.a]

dura mater In vertebrates, the connective tissue containing blood vessels that surrounds the brain and spinal cord.

E

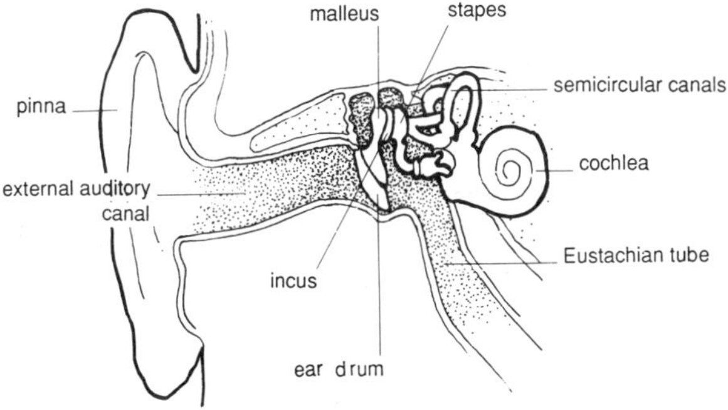

The human ear

■ **ear** One of a pair of hearing and balance sensory organs situated on each side of the head of vertebrates. In mammals it consists of three parts: the **outer**, **middle** and **inner ear**. [3.4.a] [3.5.e] [3.8.a]

■ **ear drum** Membrane at the inner end of the auditory canal of the **outer ear** which transmits sound vibrations to the **ear ossicles** of the **middle ear**. Alternative name: tympanum. [3.4.a] [3.5.e] [3.8.a]

■ **ear ossicle** Small bone found in the **middle ear** of vertebrates.

In mammals there are three in each ear, which transmit sound waves from the **ear drum** to the oval window (fenestra ovalis), which vibrates fluid in the **inner ear**, causing an impulse to travel via the **auditory nerve** to the brain. The three mammalian ossicles are the **malleus, incus** and **stapes**. Amphibians, reptiles and birds have only one ear ossicle, the columella auris. [3.4.a] [3.5.e] [3.8.a]

■ **ecdysis** Alternative name for **moulting** of an animal's **exoskeleton** to facilitate growth (*e.g.* as do many insects and crustaceans). *See also* **desquamation**.

echinoderm Member of the phylum **Echinodermata**.

■ **Echinodermata** Phylum of marine invertebrates that are characterized by chalky plates embedded in the skin and the possession of **tube feet**, powered by a water vascular system. Echinoderms include starfish, brittle stars, sea-cucumbers, sea-lilies and sea-urchins. [2.5.b]

■ **ecological niche** Position that a particular species occupies within an **ecosystem**. The term both describes the function of a species in terms of interactions with other species, *e.g.* feeding behaviour, and defines the physical boundaries of the environment occupied by the species. Bats, for instance, are said to occupy an airborne niche. [2.5.d] [2.7.b] [2.9.b]

■ **ecology** Study of interaction of organisms between themselves and with their physical environment. [2.5.d] [2.7.b] [5.10.a]

ecosphere Planet looked at from an ecological point of view. Earth is the only known ecosphere.

■ **ecosystem** Natural unit that contains living and non-living components (*e.g.* a **community** and its **environment**) interacting and exchanging materials, and generally balanced as a stable system; *e.g.* grassland or rainforest. [2.9.b] [2.10.a] [2.10.b]

■ **ectoderm** Outermost **germ layer** of the embryo of a **metazoan** which develops into tissues of the **epidermis**, *e.g.* skin, hair, sense organs, enamel and in lower animals the **nephridia**. Alternative name: epiblast.

■ **ectoparasite Parasite** that lives on the outside of its **host**; *e.g.* flea. [2.5.d] [2.6.a]

ectoplasm Non-granulated jelly-like outer layer of **cytoplasm** that is located below the **plasma membrane**. It is characteristic of most amoeboid animal cells, in which at its boundary with **plasmasol** it aids cytoplasmic streaming, and thus movement; *e.g.* in amoeboid protozoa and leucocytes.

■ **effector** Tissue or organ that responds to a nervous stimulus (*e.g.* **endocrine gland, muscle**). [3.7.c]

■ **efferent** Leading away from, as applied to vessels, fibres and ducts leading from **organs**. *See also* **afferent**. [3.5.c] [3.7.c]

■ **egg** *See* **ovum**. [3.4.c]

egg membrane Thin protective membrane that surrounds the fertilized **ovum** of animals. It is secreted by the **oöcyte** and the **follicle** cells. *See also* **chorion**.

ego Psychological term for the aspect of personality concerned with rationality and common sense. *See also* **id**.

Elasmobranchll Class of animals that includes cartilaginous fish (sharks, skates and rays). Alternative name: **Chondrichthyes**.

■ **clcctrocardiogram** (ECG) Record of the electrical activity of the heart produced by an electrocardiograph machine. Alternative name: cardiogram. [3.8.b] [3.8.c]

■ **electrocardiograph** Machine that uses electrodes taped to the body to produce **electrocardiograms**.

electrodialysis Removal of salts from a solution (often a **colloid**) by placing the solution between two semipermeable membranes, outside which are electrodes in pure solvent.

■ **electroencephalogram** (EEG) Record of the electrical activity of the brain produced by an electroencephalograph machine.

■ **electroencephalograph** Machine that uses electrodes taped to the skull to produce **electroencephalograms**.

■ **electron microscope** Instrument that uses a beam of electrons from an electron gun to produce magnified images of extremely small objects, beyond the range of an optical microscope.

■ **electron transport** Process found mainly in **aerobic respiration** and **photosynthesis** that provides a source of energy in the form of **ATP**. Hydrogen atoms are used in this system and taken up by a hydrogen carrier, *e.g.* **FAD**; the **electrons** of the hydrogen pass along a chain of carriers which are in turn reduced and oxidized. This is coupled to the formation of ATP. The hydrogen atoms together with oxygen eventually form water. [3.7.c]

■ **embryo** *1.* In animals, organism formed after cleavage of the **zygote** before hatching or birth. A maturing embryo is often termed a **foetus**. *2.* In lower plants, structure that develops from the zygote of **bryophytes** and **pteridophytes**, and in higher plants is the **seed** before germination. [3.4.c] [3.6.c]

■ **embryology** Study of **embryos**, their formation and development. [3.4.c] [3.6.c]

embryo sac *1.* In lower plants, female **gametophyte** containing the **eggs**, synergids and polar and antipodal **nuclei**. *2.* In flowering plants, large oval cell in which egg fertilization and embryonic development occur.

■ **emulsion Colloidal** suspension of one liquid dispersed in another *(e.g.* milk).

■ **enamel** White protective calcified outer coating of the crown of the **tooth** of a vertebrate. It is produced by epidermal cells and consists almost entirely of calcium salts bound together by **keratin** fibres. It is extremely hard and takes much longer to decay in an old skull than the bone. The calcium salts are readily attacked by the acid produced by bacteria in **plaque**, causing **dental caries**. [3.2.b] [3.4.b]

encephalin One of two peptides which are natural **analgesics**, produced in the brain and released after injury. The encephalins have properties similar to **morphine**. Alternative names: endorphin, enkephalin.

■ **endemic** Describing a disease that continually occurs among people or animals in a particular region. *See also* **epidemic**; **pandemic**. [2.9.a] [2.9.b] [3.8.b]

■ **endocarp** Inner of the (usually) three layers of a fruit, which may be a hard stone, as in a **drupe**. *See also* **epicarp**; **pericarp**. [3.3.b] [3.4.d]

■ **endocrine gland** Ductless organ or discrete group of cells that synthesize **hormones** and secrete them directly into the bloodstream. Such glands include the **pituitary**, **pineal**, **thyroid**, **parathyroid** and **adrenal glands**, the **gonads** and **placenta** (in mammals), **islets of Langerhans** (in the pancreas) and parts of the **alimentary canal**. Their function is parallel to the **nervous system**, that of regulation of responses in animals, but it is much slower than a nervous response. Alternative name: ductless gland. [3.8.a] [3.9.b]

■ **endocrinology** Study of structure and function of **endocrine glands** and the roles of their **hormones** as the chemical messengers of the body. [3.8.a] [3.9.b]

■ **endoderm** Innermost **germ layer** of the **zygote** of a **metazoan**, which eventually develops into the **gastrula** wall as well as the lining of its archenteron canal and its derivatives, *e.g.* **liver** and **pancreas**. In birds and mammals it also forms the **yolk** sac and **allantois**.

■ **endodermis** Innermost layer of the **cortex** of plant tissue that surrounds the **vascular tissue**. It consists of a single layer of cells which controls the movement of water and minerals between the cortex and the **stele**.

endogenous Produced or originating within an organism.

■ **endolymph** In vertebrates, fluid that fills the cavity of the **middle**, **inner** and the **semicircular canals** of the **ear**. [14.6.b]

■ **endoparasite** **Parasite** that lives inside the body of its **host**, *e.g* fluke, malaria parasite, tapeworm. [2.5.d]

endoplasm Central portion of **cytoplasm**, surrounded by the **ectoplasm** and containing **organelles**. Alternative name: plasmasol.

■ **endoplasmic reticulum** (ER) Structure that occurs in cells in the form of a flattened membrane-bound sac of cell **organelles**, continuous with the outer **nuclear membrane**. When covered with **ribosomes** it is termed rough ER; in their absence smooth ER. Its main function is the synthesis of **proteins** and their transport within or to the outside of the cell. In liver cells ER is involved in detoxification processes and in **lipid** and **cholesterol** metabolism. In association with the **Golgi apparatus**, ER is involved in **lysosome** production. [3.5.a]

endorphin One of a group of **peptides** that are produced by the **pituitary gland** and which act as painkillers in the body.

■ **endoscope** Tubular optical device, perhaps using **fibre optics**, that is inserted into a natural orifice or a surgical incision to study organs and tissues inside the body.

■ **endoskeleton Skeleton** that lies inside the body of an organism, *e.g.* the bony skeleton of **vertebrates**. *See also* **exoskeleton**. [3.3.a]

■ **endosperm** Food-storage **tissue** that surrounds the developing **embryo** in monocotyledonous seed plants, providing nourishment. [3.4.c]

■ **endospore** Tough asexual **spore** that is formed by some bacteria and some fungi to resist adverse conditions.

endothelium Tissue formed from a single layer of cells found lining spaces and tubes within the body, *e.g.* lining the **heart** in vertebrates. *See also* **epithelium**.

energy transfer Process by which chemical energy in the form of food is converted to heat energy and other forms in living organisms. *See* **biomass**; **food chain**; **pyramid of numbers**; **trophic level**. [3.6.b] [3.7.a] [3.9.a]

enrichment In microbiology, isolation of a particular type of organism by enhancing its growth over other organisms in a mixed population.

■ **enterokinase Enzyme** in blood that helps to bring about clotting.

■ **environment** All the conditions in which an organism lives, including the amount of light, acidity (pH), temperature, water supply and presence of other (competing) organisms. [2.5.c]

■ **enzyme Protein** that acts as a **catalyst** for the chemical reactions that occur in living systems. Without such a catalyst most of the reactions of **metabolism** would not occur under the conditions that prevail. Most enzymes are specific to a particular **substrate** (and therefore a particular reaction) and act by activating the substrate and binding to it, although the enzyme (like all catalysts) does not form part of the final products of the reaction. *See also* **coenzyme**. [3.7.b]

■ **ephemeral** *1.* In botany, plant with a short life-cycle; germination to seed-production may occur several times in one year. *See also* **annual**; **biennial**; **perennial**. *2.* In zoology, animal with a short life-cycle; a member of the insect order Ephemeroptera (*e.g.* mayflies). [3.6.b] [3.9.b]

epiblast Alternative name for **ectoderm**.

■ **epicarp** The outer of the (usually) three layers of a fruit, which may be a skin, rind or hard shell. *See also* **endocarp**; **pericarp**. [3.4.d]

■ **epidemic** Describing a disease that, for a limited time, affects many people or animals in a particular region. *See also* **endemic**; **pandemic**. [2.9.a] [2.9.b] [3.8.b]

■ **epidemiology** Study of diseases as they affect the population, including their incidence and prevention. [2.9.a] [2.9.b] [3.8.b]

■ **epidermis** Layer of cells at the surface of a plant or animal. In plants and some invertebrates, it forms a single protective layer, often overlaid by a **cuticle** which is impermeable to water. In vertebrates, it forms the skin and is composed of several layers of cells, the outermost becoming keratinized (*see* **keratinization**). [3.4.a] [3.5.e] [3.8.a]

■ **epididymis** Long coiled tube in the **testes** of some vertebrates through which **sperm** from the **seminiferous tubules** pass, before going into the **vas deferens** and to the exterior.

epigamic Describing animal characteristics that are attractive to the opposite sex during courtship; *e.g.* in birds, the colour of feathers and types of songs.

■ **epigeal** *1.* Describing a type of seed germination in which the **cotyledons** appear above the ground. *See also* **hypogeal**. *2.* Describing animals that live above ground, as opposed to underground. [3.4.d]

■ **epiglottis** Valve-like flap of **cartilage** in mammals that closes the opening into the **larynx**, the glottis, during swallowing. [3.4.a] [3.5.e] [3.8.a]

epinephrine Alternative name for **adrenalin**.

■ **epiphysis** *1.* Growing end of a bone, at which cartilage is converted to solid bone. *2.* Alternative name for **pineal gland**.

■ **epiphyte** Plant that grows on another plant but does not feed on it (*i.e.* it is not a parasite); *e.g.* various lichens and mosses. Epiphytes use other plants for support, and absorb water from the air. *See also* **saprophyte**. [2.6.a] [3.4.d]

■ **epithelium** Animal lining **tissue** of varying complexity, whose main function is protective. It may be specialized for a particular organ; *e.g.* squamous epithelium lines capillaries and is permeable to molecules in solution, glandular epithelium contains cells that are secretory.

■ **ergosterol** White crystalline **sterol**. It occurs in animal **fat** and in some micro-organisms. In animals it is converted to vitamin D_2 by ultraviolet radiation.

■ **erosion** Gradual removal of something by natural means; *e.g.* of soil by the action of wind and rain, or of tissue from the cervix of the womb during childbirth.

■ **erythrocyte** Red **blood cell**. It contains **haemoglobin** and carries **oxygen** around the body. In mammals, erythrocytes have no **nuclei**. *See also* **leucocyte**.

■ **essential amino acid** Any **amino acid** that cannot be manufactured in some **vertebrates**, including human beings. These acids must therefore be obtained from the diet. They are as follows: arginine, histidine, isoleucine, leucine, lysine, methionine, phenylalanine, threonine, tryptophan and valine.

■ **essential fatty acid** Any **fatty acid** that is required in the diet of mammals because it cannot be synthesized. They include linoleic acid and γ-linolenic acid, obtained from plant sources. [3.7.b]

ethology Scientific study of animal behaviour in the wild.

■ **etiolation** Phenomenon that occurs in green plants grown without light, which appear yellow due to lack of formation of **chlorophyll** and are abnormally long-stemmed; the leaves also become reduced. [3.5.e] [3.6.b] [3.7.a]

eubacteria Largest group of **bacteria**, containing the most commonly encountered forms that inhabit soil and water. The group contains Gram-positive bacteria and green photosynthetic bacteria.

■ **eugenics** Theory and practice of improving the human race through genetic principles. This can range from the generally discredited idea of selective breeding programmes to counselling of parents who may be carriers of harmful genes.

eukaryote Cell with a certain level of complexity. Eukaryotes have a **nucleus** separated from the **cytoplasm** by a nuclear membrane. Genetic material is carried on **chromosomes** consisting of **DNA** associated with **protein**. The cell contains membrane-bounded organelles, *e.g.* **mitochrondria** and **chloroplasts**. All organisms are eukaryotic except for **bacteria** and **cyanophytes**, which are **prokaryotes**. Alternative name: eucaryote.

eukaryotic Describing or relating to a **eukaryote**.

■ **Eustachian tube** Channel that connects the **middle ear** with the **pharynx** at the back of the throat in mammals and some other vertebrates. It ensures that the air pressure on each side of the ear drum is equal. It was named after the Italian anatomist Bartolomeo Eustachio (1520–74). [14.6.b]

Eutheria Placental mammals, a subclass that includes all mammals except the **monotremes** and **marsupials**.

eutrophic Describing a lake or other body of water that is well supplied with nutrients. *See also* **oligotrophic**.

■ **eutrophication** Large increase in nutrients in lakes and other fresh water, leading to overgrowth of **algae** and other plants, with consequent decrease of oxygen and depletion of fish stocks. It may be caused by run-off of agricultural fertilizers or by pollution. [2.8.a]

■ **evaluation** Review of *e.g.* an experiment to try to improve on the method used so that it becomes more accurate or efficient.

■ **evergreen** Plant that possesses leaves throughout all the seasons, *e.g.* pines and firs. The leaves are shed, but only after several years and then not all at once. *See also* **deciduous**. [2.1.a] [2.3.a]

■ **evolution** Successive altering of **species** through time. Evolutionary theory states that the origin of all species is through evolution, and thus they are related by descent. *See also* **Darwinism; natural selection**. [4.9.a]

■ **excretion** Removal of waste products of **metabolism**, carried out by elimination from the body or storage in insoluble form. Products of protein metabolism are the main substances liberated (in the form of **urea** or **uric acid**). The chief organs of excretion in vertebrates are the **kidneys**. [3.5.c] [3.8.a]

■ **exhalation** The action of breathing out.

■ **exocrine gland** Gland that discharges secretions into ducts, *e.g.* salivary glands. *See also* **endocrine gland**. [3.5.a] [3.7.b] [3.8.a]

exogamy Outbreeding. *See also* **inbreeding**.

exogenous Originating outside an organism, organ or cell. The term may refer to such things as substances (*e.g.* nutrients) or stimuli (*e.g.* light).

■ **exoskeleton Skeleton** located on the external part of the body; *e.g.* in arthropods the exoskeleton is impregnated with **chitin**, which gives the animal protection. *See also* **endoskeleton**. [3.3.a]

■ **exponential growth** Growth that occurs, *e.g.* in cultures of micro-organisms, in which a population of cells increases in numbers logarithmically. [3.9.a]

■ **extracellular** External to a cell; in a multicellular organism, extracellular tissue may still be within the organism. *See also* **intracellular**. [3.5.e] [3.8.a]

■ **eye** Organ for detecting light. Its structure varies among organisms; *e.g.* the ocellus found in some invertebrates is simple. The vertebrate eye is complex in comparison, as are the **compound eyes** of adult insects. [3.4.a] [3.5.e] [3.8.a] [15.6.b]

eyespot Primitive light-sensitive structure found in many unicellular organisms; it contains **carotenoid** pigment.

■ **eye tooth** Alternative name for **canine tooth**. [3.4.b] [3.5.e] [3.7.b]

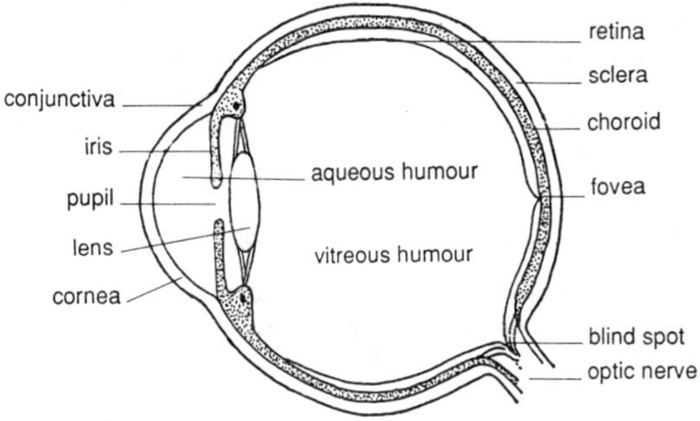

The human eye

F

facilitated diffusion Mode of transport through a **membrane** that involves carrier molecules in the membrane, which eases the transport of a specific substance but does not involve the use of energy; *e.g.* the uptake of glucose by erythrocytes (red blood cells). The transport system can become saturated with the transported substance, in contrast to simple diffusion.

FAD Abbreviation of **flavin adenine dinucleotide**.

■ **faeces** Undigested food that is eliminated from the **alimentary canal** via the **anus** after water and useful salts have been absorbed by the **colon**. [3.5.c] [3.7.b]

■ **Fallopian tube** One of a pair of tubes that in female mammals conducts **ova** (eggs) from an **ovary** to the **uterus** (womb) by ciliary action. **Fertilization** can occur when **sperm** meet eggs in the tube. It was named after the Italian anatomist Gabriel Fallopius (1523–62). Alternative name: oviduct. [3.4.c]

■ **false fruit** Fruit in which the structure is formed from an enlarged **receptacle**; *e.g.* strawberry. Alternative name: pseudocarp. [3.4.d]

false pregnancy Alternative name for **pseudopregnancy**.

■ **family** *1.* In biological **classification**, one of the groups into which an order is divided, and which is itself divided into genera; *e.g.* Canidae (dogs). *2.* A group consisting typically of parents and their children or offspring; *e.g.* a herd of elephants or pride of lions often consists of one family group. [2.3.b] [2.4.a] [2.5.b]

farina *1.* Starch or flour. *2.* Alternative name for **pollen.**

fascicle Alternative name for **vascular bundle**.

■ **fat** *See* **fats and oils**. [3.7.b]

■ **fatigue** In biology, inability of an organ or organism to function to its full capacity that results from overactivity. [3.6.d] [3.8.a]

■ **fats and oils** Naturally occurring **esters** (of **glycerol** and **fatty acids**) that are used as energy-storage compounds by animals and plants. They are hydrocarbons and members of a larger class of naturally occurring compounds called **lipids**. [3.7.b]

■ **fatty acid** Essential constituent of **fats and oils**. The simplest fatty acids are formic (methanoic) acid, HCOOH, and acetic (ethanoic) acid, CH_3COOH.

■ **fatty degeneration** Disease of tissue caused by poisoning or lack of oxygen, in which droplets of fat form within cells. [3.7.b] [3.8.b]

■ **feather** Outgrowth of a bird's skin made of **keratin**. There are various types, including contour feathers, flight feathers and down, which traps air and acts as heat insulation. [2.3.b]

■ **feature** Noticeable external aspect of an organism's appearance.

■ **Fehling's solution** Test reagent consisting of two parts: a solution of copper(II) sulphate, and a solution of potassium sodium tartrate and sodium hydroxide. When the two solutions are mixed, an alkaline solution of a soluble copper(II) complex is formed. In the presence of an **aldehyde** or **reducing sugar**, a pink-red precipitate of copper(I) oxide forms. It was named after the German chemist Hermann Fehling (1812–85). *See also* **Fehling's test**. [3.7.b]

■ **Fehling's test** Test for an **aldehyde** group or **reducing sugar**, indicated by the formation of copper(I) oxide as a pink-red precipitate with **Fehling's solution**. *See also* **Benedict's test**.

■ **femur** *1.* In four-legged (tetrapod) vertebrates, the thigh bone. *2.* In insects, the segment of the leg nearest the body. [3.3.a] [3.7.a]

■ **fermentation** Energy-producing breakdown of organic compounds by micro-organisms in the absence of oxygen; *e.g.* the breakdown of **sugar** by **yeasts** into ethanol, carbon dioxide and organic acids. Fermentation is a type of **anaerobic respiration**. [3.6.e] [7.5.b]

■ **fertilization** Fusion of specialized sex cells or **gametes** which are **haploid** to form a single cell, a **diploid zygote**. It occurs in **sexual reproduction**; *e.g.* in vertebrates the **ovum** (female gamete) is fertilized by the **sperm** (male gamete). Fertilization may be internal or external. *See also* **pollination**. [3.4.c]

■ **fertilizer** Substance used to increase the fertility of soil. Natural, or organic, fertilizers consist of animal or plant residues, which act slowly because they have to be broken down by soil bacteria before salts become soluble and available as plant food. They are usually called manures. Artificial, or inorganic, fertilizers supply nitrogen, in the form of compounds such as ammonium nitrate, ammonium sulphate and ammonium phosphate, and sometimes also phosphorus and potassium (potash). They are specified in terms of their NPK (nitrogen, phosphorus, potassium) content. [4.9.c] [5.2.a] [5.4.a] [5.6.a] [5.6.c]

■ **fibre** Thread-like structure. Natural fibres include wool and other **protein**-containing animal products, and plant fibres (*e.g.* cotton) consisting mainly of **cellulose**. The cellulose fibres in food, often referred to as roughage, are an important part of a **balanced diet**.

■ **fibrin** Meshwork of fibres formed from **fibrinogen,** which creates a **blood** clot in a wound where blood is exposed to air.

■ **fibrinogen** Soluble **plasma protein** found in blood which, after triggering of chemical factors when platelets in blood are exposed to air by wounding, is converted to **fibrin** as part of the blood-clotting mechanism. [3.5.c] [3.5.d]

fibroblast Long flat cell found in **connective tissue** which secretes **protein**; *e.g.* **collagen** and elastic fibres.

■ **fibula** Leg bone, located below the knee and outside the **tibia** (shin bone) in the hind-limb of a four-legged (tetrapod) vertebrate. [3.3.a] [3.4.a] [3.5.e]

■ **filament** *1.* The stalk of a **stamen**, which has an anther at its end. *2.* **Hypha** of a **fungus**. *3.* String of cells that make up certain **algae**. [3.4.d]

filoplume Small hair-like feathers which lack **vanes** and are scattered over a bird's body between the **contour** feathers. Filoplumes are concerned with the fine control of flight direction.

■ **filter feeding** Acquisition of nutrients, characteristic of non-motile aquatic organisms, that involves straining out small particles of organic matter suspended in water.

■ **filtration** Method of separating solid particles from a mixture of the particles in a liquid by straining it through a porous material (such as filter paper or glass wool), through which only the liquid passes. The process can be accelerated using suction.

■ **fin** Locomotory appendage on a fish which provides a large surface area for steering and swimming.

■ **fish** Aquatic vertebrate animals that are usually divided into three classes: the **Agnatha** (jawless fish, *e.g.* lampreys), the **Elasmobranchii** (cartilaginous fish, also called Chondrichthyes, *e.g.* sharks and rays) and the **Osteichthyes** (bony fish, *e.g.* cod, salmon, etc.). [2.3.b] [2.4.a] [2.5.b]

■ **fission** Splitting. In biology, division of a cell or single-celled organism into two (*see* **meiosis; mitosis**). [3.5.a]

■ **fixation of nitrogen** Part of the **nitrogen cycle** that involves the conversion and eventual incorporation of atmospheric nitrogen into compounds that contain nitrogen. Nitrogen fixation in nature is carried out by nitrifying soil **bacteria** or blue-green **algae** (Cyanophytae) in the sea. Soil bacteria may exist symbiotically (*see* **symbiosis**) in the root nodules of **leguminous** plants or they may be free-living. Small amounts of nitrogen are also fixed, as nitric oxide, by the action of lightning. [2.7.a] [5.4.a] [5.5.a]

■ **flaccid** Describing tissue that has become soft and drooping, usually because of loss of water. [3.4.a] [3.5.e] [3.8.a]

■ **flagellate** Single-celled **protozoan** animal that moves by beating flagella (*see* **flagellum**). [2.4.a] [3.4.c]

■ **flagellum** Long hair-like **organelle** whose beating movement causes locomotion or the movement of fluid over a cell. Flagella are present in most motile **gametes** and unicellular plants or animals (*e.g.* protozoa), in which they occur singly or in small clusters. In some multicellular organisms (*e.g.* sponges and hydra) they are used for circulation of water containing food and respiratory gases. *See also* **cilium**.

■ **flame cell** Excretory cell possessed by some invertebrates. Waste products are drawn in and moved outside by **cilia**. Alternative name: solenocyte.

flavin adenine dinucleotide (FAD) **Coenzyme** that functions in the oxidation-reduction reactions of **enzymes**, *e.g.* the oxidative degradation of **pyruvate, fatty acids** and **amino acids**, and in **electron transport**. Alternative name: flavine.

flavonoid Aromatic, oxygen-containing **heterocyclic** organic compound. Many natural pigments are flavonoids.

flavoprotein Member of a group of conjugated **proteins** in which the **prosthetic group** constitutes a derivative of **riboflavin** (*e.g.* FAD or FMN). Flavoprotein dehydrogenases (enzymes) are involved in the **electron transport** chain of **aerobic respiration**.

■ **flocculation Coagulation** of a finely divided precipitate into larger particles. *E.g.* in farming, flocculation of clay is deliberately brought about by the addition of lime, thus improving the drainage of clay soils. [2.9.b] [3.7.a] [5.6.a]

■ **flower** In flowering plants (**angiosperms**), the organ of **sexual reproduction**, including the male **stamens** and female **carpels**. [3.1.a] [3.4.d]

fluke *1.* Any **endoparasitic** flatworm that belongs to the class Trematoda. *See also* **bilharzia**. *2.* Tail of a whale, dolphin or porpoise.

■ **fluoridation** Addition of inorganic fluorides to drinking water to combat dental decay. [5.6.a] [5.7.a]

■ **fluorocarbon** Very stable organic compound in which some or all of the hydrogen atoms have been replaced by fluorine. Fluorocarbons are used as solvents, aerosol propellants and refrigerants. Their use is being limited because they have been implicated in damage to the **ozone layer** of the atmosphere. [5.5.a]

■ **fluoroscope** Fluorescent screen that allows direct observation of **X-ray** images, often connected to a camera. It is used in medicine (radiography) and industrial X-ray applications.

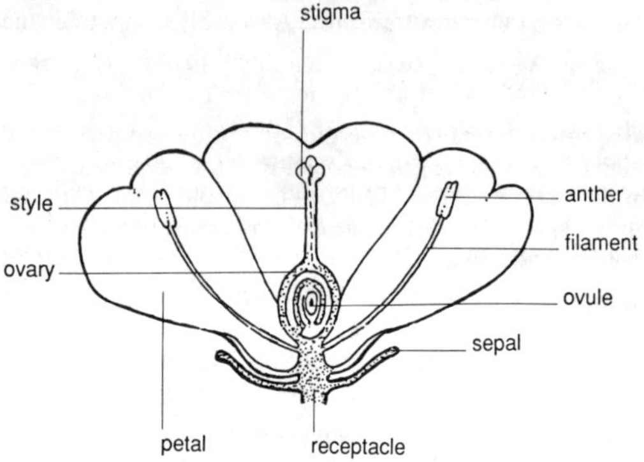

Structure of a flower

■ **focusing** Adjustment of an optical device so that it produces a sharp image. In the **eye**, for instance, ciliary muscles alter the shape of the lens to produce a clear image of an object on the retina.

■ **foetus** In mammals, an **embryo** after a certain stage of development, usually when it begins to resemble the developed animal (in human beings after about two or three months of pregnancy). [3.4.c]

■ **folic acid** Water-soluble B-group **vitamin**. Its deficiency leads to **anaemia** and it is important in the formation of various **coenzymes**, which are in turn essential for growth and reproduction of cells.

■ **follicle** *1.* In botany, dry dehiscent fruit formed from a monocarpellary (with a single **carpel) ovary** which at maturity splits along one edge to release its seeds; *e.g.* columbine fruit. *2.* Cavity, sac or gland within an organ or tissue; *e.g.* **Graafian follicle**, hair follicle. [3.8.a]

■ **follicle-stimulating hormone** (FSH) Member of a group of **hormones** that are secreted by the anterior lobe of the **pituitary gland** in vertebrates. It stimulates the growth and maturation of **ovarian follicles** and the growth only of oöcytes, which are matured under the action of **luteinizing hormone**. In males FSH stimulates **sperm** formation in the **testes**. [3.8.a]

food additive *See* **additive**. [3.5.c] [3.7.b]

■ **food chain** Food relationship between organisms in an **ecosystem** in which energy is transferred from plants, the producers, through a series of organisms, the consumers. Each stage of the food chain is a trophic level. The first level is occupied by plants, which obtain their energy from the Sun. The second trophic level is occupied by **herbivores** (plant-eating animals), which are in turn eaten by the **carnivores** (meat-eating animals). Food chains in a **community** are interconnected because most organisms consume more than one type of food, thus forming a more complicated **food web** or food cycle. [2.7.b]

■ **food consumer** *See* **food chain**.

■ **food producer** *See* **food chain**.

■ **food web** Complex relationship formed between organisms due to mode of nutrition, involving a network of **food chains** in an environment.

■ **forebrain** Largest and topmost portion of the vertebrate brain that comprises the **cerebral hemispheres** and the **basal ganglion**. [3.7.c]

■ **formalin** 40% solution of formaldehyde (methanal) in water, which used to be employed for preserving biological specimens.

■ **fossil** Remains, impressions or casts of dead animals and plants preserved in rocks. Because organic matter rots away quickly, a fossil usually consists of skeletal material which is partly or wholly replaced by mineral deposits from circulating water. Burrows, footprints or faeces may also become fossilized. [2.3.b] [2.4.a] [2.5.b]

■ **fovea** Part of the **retina** of the **eye** that has a concentration of **cones** (but no **rods**), which comes into play when acute vision is required. Alternative name: yellow spot. [15.6.b]

fraternal twins *See* **twins**. [3.4.c]

■ **freeze drying** Method of drying a heat-sensitive substance such as blood plasma or food by freezing it below 0°C and then removing the frozen water by volatilization in a vacuum.

fructification Process of forming a **fruiting body**. The term also sometimes refers to the fruiting body itself.

■ **fructose** $C_6H_{12}O_6$ Fruit sugar, a **monosaccharide** carbohydrate (**hexose**) found in sweet fruits and honey. Alternative name: laevulose. [3.7.b]

■ **fruit** Plant tissue that develops from the **ovary** of a flowering plant and forms around the maturing seeds, following **pollination** and subsequent **fertilization**. The term is also sometimes used to describe the mature seeds of some coniferous trees, *e.g.* juniper 'berries'. *See also* **berry**; **capsule**; **drupe**; **false fruit**; **follicle**.

■ **fruiting body** Structure that is developed by many **fungi** in which **spores** are formed. [3.4.d]

■ **FSH** Abbreviation of **follicle-stimulating hormone**.

fucoxanthin Brown **carotenoid** plant pigment, present in brown **algae**.

■ **fungicide** Substance that kills **fungi**. [2.9.a] [2.9.b] [2.9.c]

■ **fungus** Mainly terrestrial plant-like organism, different from other plants because of its lack of **chlorophyll**. Most fungi are **saprophytic** or **parasitic** organisms (*e.g.* moulds) whose walls consist of **chitin**, although a few produce cellulose as well. They are classified as the plant division (phylum) Mycota, although some authorities put them in a kingdom of their own. [2.3.b] [2.4.a] [2.5.b] [2.9.a] [2.9.b] [2.9.c]

funicle Stalk that connects the **ovule** to the **placenta** in the ovaries of **angiosperm** plants. [3.4.d]

G

■ **galactose** $C_6H_{12}O_6$ **Monosaccharide** sugar that occurs in milk and in certain gums and seaweeds as the **polysaccharide** galactan.

■ **gall** *1*. In botany, swelling on a plant, usually caused by a parasite. *2*. In zoology, alternative name for **bile**. [3.4.a] [3.5.e] [3.7.b] [3.8.a]

■ **gall bladder** Storage organ in some vertebrates that, stimulated by hormones, releases **bile** (along the bile duct) to the **duodenum** during **digestion**. [3.4.a] [3.5.e] [3.7.b] [3.8.a]

■ **gallstone** Accretion, usually of **cholesterol** or calcium salts, that occurs in the **gall bladder** or its ducts. Alternative name: biliary calculus. [3.4.a] [3.5.e] [3.7.b] [3.8.a]

gametangium In plants, an organ in which **gametes** are produced.

■ **gamete** Specialized **sex cell** (*e.g.* an ovum or a sperm in animals or an ovule or pollen in plants), which is **haploid** – *i.e.* containing half the normal number of **chromosomes**. Gametes combine at **fertilization** to form a **zygote** that develops into a new organism (with the normal, **diploid**, chromosome number). *See also* **parthenogenesis**. [3.4.c]

■ **gametophyte** Stage in the life-cycle of certain plants, especially mosses and ferns, which show **alternation of generations**. It produces **gametes** and is **haploid** (as distinct from the following **sporophyte** stage, which is **diploid**).

gamma globulin Alternative term for **immunoglobulin**.

■ **ganglion** Area of nervous tissue that contains a complex set of **synapses**. In vertebrates, ganglia constitute much of the **central nervous system** and occur in the **sympathetic** and **parasympathetic** nervous systems, but differ in their structure in each case. [3.7.c]

■ **gas exchange** Part of **respiration** in which organisms exchange gases (carbon dioxide and oxygen) with their environment: air for terrestrial plants and animals, water for aquatic ones. It may involve the use of **lungs** (mammals, birds, adult amphibians and reptiles), **gills** (larval amphibians, fish and other aquatic animals), **spiracles** (insects and other terrestrial arthropods) or **stomata** (green plants). In other organisms (*e.g.* aquatic plants, fungi) gas exchange takes place directly between cells and the environment. [3.5.e] [3.6.a]

■ **gastric** Relating to the stomach or digestion. [3.5.c] [3.5.e] [3.7.b]

■ **gastric juice** Fluid secreted by glands in the stomach wall during **digestion**; it contains two principal **enzymes**, **pepsin** and **rennin**, and hydrochloric acid. [3.5.c] [3.5.e] [3.7.b]

gastropod Member of the mollusc class **Gastropoda**.

Gastropoda Class of **molluscs** characterized by a single flattened foot, a head bearing eyes on stalks and usually a coiled shell. The gastropods include slugs (which have no external shells), snails and whelks.

gastrula Cup-shaped structure in early embryonic development, the stage that succeeds the **blastula**.

■ **gel** Jelly-like colloidal solution. *See* **colloid**.

■ **gelatin** Protein extracted from animal hides, skins and bones, which forms a stiff jelly when dissolved in water. It is used as a clarifying agent, in foodstuffs, and in the manufacture of adhesives. Alternative name: gelatine.

gemmation *See* **budding**.

■ **gene** Segment of **DNA** that specifies a complete **polypeptide** chain, which can be regarded as an independent inheritable unit. Genes specify particular traits and are passed from parent to offspring. Some genes may be **dominant** to others that may be **recessive**. Different forms of a gene are **alleles**. They may undergo **mutation** to give different characteristics. Every cell of an organism has a set of genes on its **chromosomes** that are identical to those in every other cell, having ultimately arisen from mitotic division of a **zygote**. [4.5.a] [4.9.d]

■ **gene pool** Total of all the different **genes** in a population.

■ **genera** Plural of **genus**.

generation time Time within a population of cells (*e.g.* micro-organisms) that it takes them to undergo division to form pairs of **daughter cells;** *i.e.* the doubling time.

gene therapy Treatment of disorders by altering the genetic material in a cell; *e.g.* by microinjection of favourable **genes** into a **germ cell**.

■ **genetic code** Code possessed by **DNA** which contains instructions for **protein synthesis** in a **cell**. Each **amino acid** is specified by a triplet of **bases** located in the DNA, and each protein is in turn specified by a particular sequence of **amino acids**. The code thus contains instructions for all **enzymes** produced in the cell, and consequently the characteristics of the organism. [4.5.a] [4.5.d]

■ **genetic engineering** Manipulation of genetic material such as **DNA** for practical use, *e.g.* the introduction of foreign **genes** into micro-organisms for the production of a useful **protein** (such as human insulin). The technique is also used in the study of genetic material. [4.10.d]

■ **genetics** Study of inheritable characteristics of organisms – *i.e.* **heredity**.

genetic variation Differences between members of the same species resulting from slight changes in the make-up of the genes they inherit from their parents. Such differences may in turn be inherited and are significant in adaptation (and therefore species survival) in a changing environment. *See also* **variation**. [4.4.a] [4.6.a]

genome Entire genetic material present in a cell of an organism.

■ **genotype** Genetic constitution of an organism, *i.e.* the characteristics specified by its **alleles**. It is the genetic make-up of the organism, as opposed to the way its genes are expressed in its appearance (which is the **phenotype**).

■ **genus** In biological **classification**, one of the groups into which a family is divided, and which is itself divided into species; *e.g. Vulpes* (foxes). *See also* **binomial nomenclature**. [1.7.f] [2.5.b]

■ **geotaxis** Change in the direction of movement of a mobile organism that is made in response to gravity. *See also* **geotropism**.

■ **geotropism** Growth in response to **gravity** in plants. Orientation of roots and shoots differs with respect to gravity. Roots grow towards gravity and are said to be positively geotropic; shoots grow away from gravity and are negatively geotropic. The effect is caused by plant growth **hormones** or **auxins**, which affect the shoot and root differently.

■ **germ** Imprecise term for a micro-organism that can cause disease; a pathogen. Germs include **bacteria**, **protozoa** and **viruses**. [3.8.b]

■ **germ cell** Sex cell, or **gamete**.

■ **germination** Beginning of the development of an embryo in a **seed** into a plant, which commences with the growth of the **radicle** and **plumule**, and is initiated by the right conditions of moisture, oxygen supply and temperature. *See also* **epigeal**; **hypogeal**. [3.4.d] [3.7.a]

■ **germ layer** Layer of cells present in an **embryo**. In **triploblastic** organisms the layers consist of **ectoderm**, **mesoderm** and **endoderm**, which each give rise to particular **tissues**.

■ **gestation** Time of development of an **embryo** from fertilization to birth in **viviparous** animals; the duration of **pregnancy**. [3.7.c]

gibberellin Any one of a group of plant **hormones** that stimulate rapid growth.

■ **gill** *1*. In zoology, respiratory organ in fish and certain other aquatic animals that extracts dissolved oxygen from water. *2*. In botany, thin vertical structure under the fruiting body of a fungus that carries the spore-bearing hymenium. [3.5.e]

■ **gizzard** Part of the gut of animals that eat indigestible food. It has thick muscular walls and a tough grinding inner surface specialized for breaking up food. Gizzards of birds and earthworms contain small stones or grit; those of the **arthropods** have chitin patches and spines. Alternative name: gastric mill. [3.4.a] [3.5.e] [3.8.a]

■ **gland** *1*. In animals, organ that synthesizes and secretes specific chemicals, either directly into the bloodstream **(endocrine gland)** or through a duct **(exocrine gland)** into tubular organs or onto the body surface. *2*. In plants, specialized unicellular or multicellular structure involved in the secretion of various substances formed as by-products of plant metabolism. [3.8.a]

globulin Water-insoluble protein that is soluble in aqueous solutions of certain salts. Globulins generally contain **glycine** and are coagulated by heat; *e.g.* immunoglobulin.

■ **glomerulus** Ball of **capillaries** located in the **Bowman's capsule** of the kidney. [3.5.e] [3.8.a]

■ **glottis** Opening of the **larynx,** through which air passes from the pharynx to the **trachea** (windpipe) of vertebrates. [3.4.a] [3.5.e] [3.7.b] [3.8.a]

glucagon In animals, a polypeptide hormone that (like **insulin**) is synthesized and secreted by the **islets of Langerhans** in the **pancreas**. Produced in response to low blood pressure, it stimulates **glycogen** breakdown in the liver, with release of **glucose** into the bloodstream. Its action is therefore opposite to that of insulin (which reduces blood glucose levels).

■ **glucose** $C_6H_{12}O_6$ **Monosaccharide** carbohydrate, a soluble colourless crystalline **sugar** (hexose) which occurs abundantly in plants. It is the principal product of **photosynthesis** and source of energy in both plants and animals. It is a product of carbohydrate digestion and is the sugar in blood. Glucose is used commercially in the manufacture of confectionery and the production of beer. Its natural polymers include **cellulose** and **starch**. Alternative names: dextrose, grape sugar. [3.7.b]

glucoside *See* **glycoside**.

gluten Protein that occurs in cereals, particularly wheat flour. People with coeliac disease cannot tolerate gluten, and have to eat a gluten-free diet.

glyceride **Ester** of **glycerol** with an organic acid. The most important glycerides are **fats and oils**.

■ **glycerol** $HOCH_2CH(OH)CH_2OH$ Colourless sweet syrupy liquid, a trihydric **alcohol** that occurs as a constituent of **fats and oils** (as glycerides), from which it is obtained commercially. It is used in foodstuffs and medicines. Alternative names: glycerin, glycerine, propan-1,2,3-triol. [3.7.b]

glycine $CH_2(NH_2)CO_2H$ Simplest **amino acid**, found in many **proteins** and certain animal excretions. It is a precursor in the biological synthesis of **purines, porphyrins** and **creatine**. It is also a component of **glutathione** and the bile salt **glycocholate**. It acts as a **neurotransmitter** at inhibitory nerve **synapses** in vertebrates. Alternative names: aminoacetic acid, aminoethanoic acid.

■ **glycogen** **Polysaccharide** carbohydrate, the main energy store in the liver and muscles of vertebrates ('animal starch'), also found in some algae and fungi. **Amylase** enzymes convert it to **glucose**, for use in metabolism. [3.7.b]

glycolipid Member of a family of compounds that contain a **sugar** linked to **fatty acids**. Glycolipids are present in higher plants and neural tissue in animals. Alternative names: glycosylacylglycerols, glycosyldiacylglycerols.

glycolysis Conversion of **glucose** to lactic or pyruvic acid with the release of energy in the form of **adenosine triphosphate** (ATP). In animals it may occur during short bursts of muscular activity.

glycoprotein Member of a group of compounds that contain **proteins** attached to **carbohydrate** groups. They include **blood glycoproteins**, some **hormones** and **enzymes**.

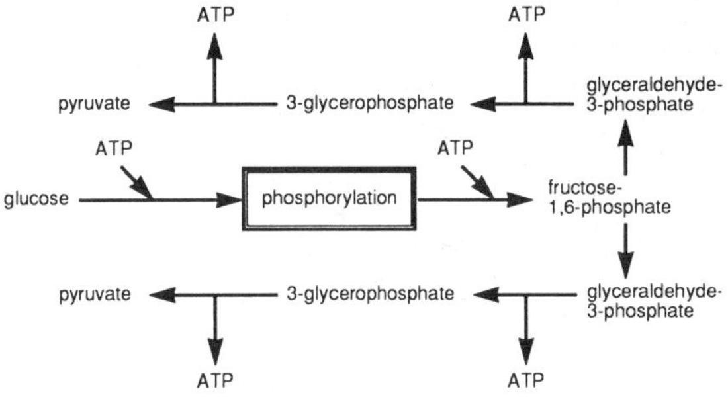

Production of ATP by glycolysis

glycoside Compound formed from a **monosaccharide** in which an alcoholic or phenolic group replaces the first hydroxyl group. If the monosaccharide is **glucose**, it is termed a glucoside.

Gnathostomata Superclass containing all vertebrates that possess jaws.

goblet cell Mucus-secreting cell in **mucous membranes**.

■ **Golgi apparatus Organelle** that occurs in most cells as layers of flattened membrane-bounded sacs. It is involved in the formation of **zymogen**, synthesis and transport of secretory **polysaccharides** (*e.g.* **cellulose** in cell plate or secondary cell-wall formation) and formation of mucus in **goblet cells**;

assembly of **glycoproteins**; packing of **hormones** in nerve cells that carry out neurosecretion; formation of **lysosomes**; and probably production of the **plasma membrane**. It was named after the Italian histologist Camillo Golgi (1843–1926). Alternative names: Golgi body, Golgi complex.

■ **gonad** Reproductive organ of an animal, *e.g.* **ovary** or **testis,** in which **ova** (eggs) and **sperm** are formed respectively. Gonads may also function as **endocrine glands,** secreting sex **hormones.** [3.4.c] [3.9.b]

■ **gonadotrophic hormone Hormone** that acts on the **gonads,** controlling the initiation of puberty, the menstrual cycle and lactation in females and sperm-formation in males. It is produced by the pituitary gland. It is also used for the treatment of infertility in women. Alternative name: gonadotrophin. [3.4.c] [3.5.e] [3.8.a]

■ **Graafian follicle** Fluid-filled ball of cells in the mammalian **ovary** inside which an **oöcyte** develops. It matures periodically and then bursts at the surface of the ovary (at **ovulation**) to release an **ovum** (egg). The follicle then temporarily becomes a solid body, the **corpus luteum**. It was named after the Dutch anatomist Regnier de Graaf (1641–73). Alternative name: ovarian follicle. [3.4.c] [3.8.a]

■ **graft** Transplantation of an **organ** or **tissue** from one organism into the body of another, although, apart from cornea and skin grafts, the term is applied mostly to plants (other animal tissue grafts are usually called transplants). [3.8.b]

■ **graft hybrid** Plant that is an unusual mixture of two genetically different tissues; *e.g.* grafting purple broom with laburnum produces a tree that has the shape of laburnum but with purple flowers. A graft hybrid is a form of **chimaera**. [3.8.b]

■ **grafting** Method of propagating plants using a **graft** (a stem or bud from the scion inserted into the tissues of the stock). [3.8.b]

■ **Graminae** Large family of flowering plants that consists of the grasses. There are more than 90,000 species world-wide, a very few of which (wheat, rice, maize and millet) provide the staple food of most of the world's population. The bamboos are the largest, and have woody stems. [2.3.b] [2.4.a] [2.5.b]

Gram's stain Staining technique that differentiates between and classifies two major groups of bacteria, known as Gram-positive and Gram-negative bacteria, which stain deep purple or red respectively. The different uptake of stain is due to differences in cell-wall structure. It was named after the Danish physician Hans Christian Gram (1853–1938).

■ **grape sugar** Alternative name for **glucose**.

■ **greenhouse effect** Overheating of the Earth's atmosphere resulting from atmospheric pollution, particularly the build-up of carbon dioxide (CO_2), which absorbs and thus traps some of the solar radiation reflected from the Earth's surface. [5.9.a]

■ **growth** Increase in size, dry mass or numbers that is a characteristic of living organisms.

■ **growth hormone** Any of a group of **polypeptides** that control growth and differentiation in animals and plants. In animals it is secreted by the anterior **pituitary gland** and acts directly on the cells of the body, particularly those of bones. Alternative names: (animals) somatotrophin; (plants) auxin; growth substance. [3.7.a] [3.8.a] [3.9.b]

■ **growth ring** Alternative name for **annual ring**.

■ **growth substance** Any of the substances (sometimes called plant hormones) that regulate the growth and development of plants. Growth substances include auxins and gibberellins.

guanine $C_5H_5N_5O$ Colourless crystalline organic base (a **purine** derivative) that occurs in **DNA**.

■ **guard cell** One of a pair of cells that control the opening (and closing) of **stomata,** in turn controlling the **gas exchange** and **transpiration** in a leaf. [3.6.b] [3.7.a] [3.9.a]

■ **gut** Alternative name for **alimentary canal**. [3.4.a] [3.5.e] [3.8.a]

gymnosperm Member of the **Gymnospermae**.

Gymnospermae Subdivision of the plant division (phylum) **Spermatophyta**. Gymnosperms are cone-bearing plants (typically cycads and conifers) whose seeds have an **ovule** that is not enclosed by the **carpel**.

■ **gynaecium** Female reproductive part of a flower. Alternative name: carpel. [3.4.d]

H

■ **habitat** Part of the **environment** in which an organism or a community of plants and animals lives; *e.g.* a lake or a forest. [2.2.a] [2.3.c] [2.5.a]

habituation Process by which the nervous system becomes accustomed to a particular **stimulus** and after a time is no longer irritated by it; *e.g.* the feel of one's clothes or the background noise of machinery.

■ **haem** Iron-containing group of atoms attached to a **polypeptide** chain; *e.g.* in **haemoglobin** and **myoglobin**.

haemocyanin Blue **blood** pigment that contains copper as its prosthetic group for the transport of oxygen. It is confined to lower animals, *e.g.* molluscs.

■ **haemoglobin** Red **blood** pigment that contains four iron **haem** prosthetic groups for the transport of oxygen. It occurs in red blood cells (erythrocytes) of vertebrates and some invertebrates. [3.5.b] [3.8.a]

haemolysis Breakdown of red blood cells (erythrocytes) which results in the release of **haemoglobin**.

■ **haemophilia** Inherited disease that affects human males in which there is a deficiency in the blood-clotting process. Sufferers are at constant risk of bleeding excessively from even minor injuries.

■ **hair** *1.* Derivative of **ectoderm** composed of insoluble **proteins** or **keratins**. Its role in mammals includes assisting regulation of body temperature. *2.* In plants, any of various outgrowths from the epidermis, *e.g.* root hairs, which absorb water.

halophilic Exhibiting a preference for an environment containing salt (*e.g.* sea-water). The term is usually applied to **bacteria**.

■ **halophyte** Plant that lives in a salty environment. [2.5.a]

■ **haploid** Having half the number of **chromosomes** of the organism in the cell **nucleus**. The haploid state is found in **gametes** and in the **gametophytes** of plants that exhibit **alternation of generations**. *See also* **diploid**. [4.6.a] [4.8.a]

Haversian canal Any of the numerous channels that occur in **bone** tissue, containing blood vessels and nerves. An organic matrix is laid down in layers encircling each Haversian canal. It was named after the English physician Clopton Havers (? – 1702).

■ **hay fever** Excessive secretion of mucus from the nasal passages caused by the inhalation of an **antigen** such as pollen or fungal spores. It is a type of **allergy**. [3.8.a] [3.8.b]

■ **heart** Muscular organ in vertebrates that pumps **blood** into a system of **arteries**. Blood returns to the heart from the tissues via **veins**, and is passed to the **lungs** to become reoxygenated. The mammalian heart is divided into four chambers: the right and left atria (also known as auricles) and right and left, thick-walled ventricles, so that deoxygenated and oxygenated blood remain separate. [3.4.a] [3.5.e] [3.8.a]

■ **heartbeat** Alternate contraction and relaxation of the **heart**, corresponding to **diastole** and **systole**.

Hemichordata Subdivision of the **Chordata** (consisting of the acorn worms). Hemichordates lack a true **notochord** and have a very primitive nervous system.

hemichordate Member of the **Hemichordata**.

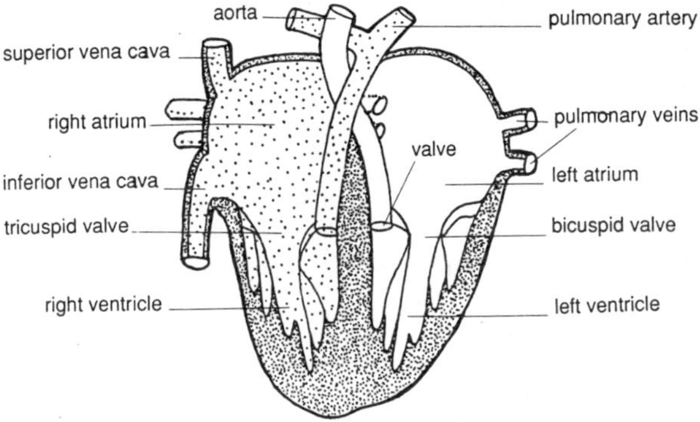

The human heart

Hemiptera Order of winged sap-sucking insects, often referred to as the true bugs. Hemipterans include aphids and leaf-hoppers.

■ **heparin Polysaccharide** substance that prevents the clotting of **blood** by inhibition of the conversion of prothrombin to **thrombin**; used medicinally as an anticoagulant. [3.8.a] [3.8.b]

■ **hepatic** Relating to the **liver**. [3.4.a] [3.5.e] [3.7.b] [3.8.c]

■ **hepatic portal system** In vertebrates, system of blood **capillaries** into which dissolved foods (except for **fatty acids** and **glycerol**, which enter the **lymphatic system**) pass from the intestine lining for transport to the liver. [3.4.a] [3.5.e] [3.7.b] [3.8.c]

■ **herbicide** Chemical that kills plants; a weed-killer. Herbicides include defoliants, contact herbicides (which kill only the parts they touch) and selective herbicides (which kill only certain plants). [2.9.c] [4.9.c]

■ **herbivore** Animal that feeds on plants, with teeth and a digestive system adapted for that purpose; *e.g.* cattle, deer, goats, rabbits, sheep. *See also* **carnivore**; **omnivore**. [2.3.b] [2.4.a] [2.5.b]

■ **heredity** Mechanism by which offspring inherit certain characteristics from their parents. *See* **genetics**. [4.9.c]

■ **hermaphrodite** Plant or animal that possesses both male and female organs. Alternative name: bisexual (particularly of plants). [2.6.a] [3.4.c] [3.4.d]

■ **herpes virus** Any of a group of animal DNA **viruses** responsible for various diseases; *e.g.* chickenpox (variola) and herpes simplex (coldsore), and herpes zoster (shingles). [3.8.b]

■ **heterogametic** Describing an organism that produces two kinds of **gametes**, each possessing a different **sex chromosome**. These are usually produced by the male; *e.g.* in human males half the **sperms** contain an **X-chromosome** and half a **Y-chromosome**. [2.6.a]

■ **heterotrophism** Mode of nutrition exhibited by most animals, fungi, bacteria and some flowering plants. It involves the intake of organic substances from the **environment** due to the inability of the organism to synthesize them from inorganic materials. *See also* **autotrophism**; **holophytic**. [2.5.a] [3.7.a]

■ **heterozygous** Describing an organism that possesses two dissimilar **alleles** in a pair of **chromosomes**. A **dominant** allele can be expressed in the heterozygous or **homozygous** state, but a **recessive** allele can be expressed only in the homozygous state. [4.6.a]

■ **hexose** **Monosaccharide** carbohydrate (sugar) that contains six carbon atoms and has the general formula $C_6H_{12}O_6$; *e.g.* **glucose** and **fructose**.

■ **hibernation** Winter dormancy that occurs in some animals. It involves the slowing of **metabolism** and a drop in body temperature, and provides a means of avoiding the necessity to maintain a high body temperature during winter. *See also* **aestivation**. [2.3.c]

hinge joint Simple form of articulation between bones that allows for movement in a single plane (*e.g.* elbow joint, knee joint). [3.3.a]

hippocampus *1.* Often refers to *Hippocampus*, the sea horse, a genus of fish. *2.* Structure of nervous tissue within the vertebrate **brain**, which in human beings is believed to be involved in the process of memory.

Hirudinea Class of **annelids** consisting of the leeches, some of which are blood-sucking.

Hirundinidae Family of birds with pointed wings and forked tails, which includes swallows and martins.

■ **histamine** Organic compound that is released from cells in **connective tissue** during an allergic reaction (*see* **allergy**). It causes dilation of **capillaries** and constriction of **bronchi**. [3.8.a]

histidine $(C_3H_3N_2)CH_2CH(NH_2)COOH$ Crystalline soluble solid, an optically active **essential amino acid**. Alternative name: 2-amino-3-imidazolylpropanoic acid.

■ **histocompatibility lymphocyte-A system** (HL-A system) In animals, genetically determined **glycoprotein antigens** that are secreted by the **macrophages** and **lymphocytes** in the cell membrane. They show a strong immunological response, leading to rejection of transplanted tissue or organs. Alternative name: histocompatibility antigens.

histogram Type of **bar chart** in which the area of a bar or block represents a quantity.

■ **histology** Scientific study of the structure of **tissues**.

histone One of a group of small **proteins** with a large proportion of basic **amino acids**, *e.g.* arginine or lycine. Histones are found in combination with **nucleic acid** in the **chromatin** of **eukaryotic** cells.

HL-A system Abbreviation of **histocompatibility lymphocyte-A system**.

holoenzyme Enzyme that forms from the combination of a **coenzyme** and an **apoenzyme**. The former determines the nature and the latter the specificity of a reaction.

■ **holophytic** Describing an organism that lives by **photosynthesis**; *e.g.* most green plants. *See also* **autotrophism**. [2.5.a]

Holothuroidea Class of **Echinodermata**, elongated marine invertebrates that may be free-living or attached to a surface; the sea-cucumbers.

■ **holozoic** Describing organisms that feed on solid organic matter or other organisms; *e.g.* most animals and insect-eating plants. *See also* **heterotrophism**. [2.5.a]

■ **homeostasis** Maintenance of a constant chemical and physical state in the internal environment of a cell, organ or organism, *e.g.* maintenance of body temperature or the balance of salts in the blood. [3.8.a] [3.8.b] [3.10.a]

■ **homoeopathy** System of alternative medicine that treats disorders by introducing substances into the body that provoke similar symptoms and so encourage the body's own defences. Medical opinion on the value of homoeopathy is divided. Alternative name: homeopathy.

■ **homogametic** Describing an organism with homologous **sex chromosomes** (*e.g.* XX), and therefore producing **gametes** each containing one (X) chromosome. In human beings and many other mammals, females are the homogametic sex.

■ **homogamy** In flowering plants, maturation of the **anther** and **stigma** at the same time.

■ **homogeneous** In biology, describing similar structures found in different species that are thought to have originated from a common ancestor.

■ **homoiothermic** Describing animals that maintain a more or less constant body temperature (*e.g.* mammals, birds). Alternative name: warm-blooded. [3.8.a]

■ **homologous** Describing things with common origin, but not necessarily the same appearance or function (*e.g.* the arms of a human and the wings of a bat). *See also* **analogous**.

■ **homologous chromosome Chromosome** that undergoes pairing during **meiosis**. Each one of the pair carries the same **genes** but not always the same **alleles** for a given character as the other member of the pair. Thus they have the same **gene** loci, **centromeres** at the same points and are very similar in length and shape. At **fertilization** each parent contributes one homologue of each pair. [4.5.a]

■ **homologue** In biology, a **homologous chromosome**.

■ **homozygous** Having the same two **alleles** for a given **gene**. Two organisms that are homozygous for the same alleles breed true for the character in question, thus producing progeny which are homozygous and identical to the parent with respect to that gene. *See also* **heterozygous**. [4.5.a]

■ **hormone** *1.* In animals, chemical 'messenger' of the body that is secreted directly into the bloodstream by an **endocrine gland**. Each gland secretes hormones of different composition and purpose which exert specific effects on certain target tissues. *2.* In plants, organic substance that at very low concentrations affects growth and development; *e.g.* auxin, gibberellin, abscisin, florigen. *See* **growth substance**. [3.8.a] [3.9.b]

■ **horn** Hard structure generally made of **keratin** usually borne in pairs on the heads of certain animals (*e.g.* most antelopes, cattle and deer). Antlers (horns of deer) are shed and regrown every year.

horticulture Commercial growing of plants for gardens and parks. *See also* **agriculture**.

■ **host** *1.* Organism on which a parasite lives (*see* **parasitism**). *2.* Tree on which an **epiphyte** grows. [2.5.d] [2.6.a]

■ **humerus** Upper bone in the forelimb of a tetrapod vertebrate; in human beings, the upper arm bone. [3.3.a] [3.4.a] [3.5.e]

■ **humus** *1.* Dark brown organic material produced in soil by the action of **decomposers** on dead plant and animal matter. *2.* Topmost layer of soil that contains a high proportion of decomposed organic material, known also as leaf litter. [5.2.a] [5.4.a]

■ **hybrid** Progeny of plants or animals produced from the cross of two genetically dissimilar parents. The term is usually limited in use to offspring of parents from two different (but closely related) species; *e.g.* a hybrid of a mare (female horse) and a male donkey is a mule.

■ **hybridization** Crossing of animals or plants to produce a **hybrid**.

■ **hybrid vigour** General improvement of physical characteristics of a **hybrid** in comparison to its parents resulting from an increase in genetical variation.

■ **hydra** Small freshwater animal of the phylum **Cnidaria**.

■ **hydrocarbon** Organic compound that contains only carbon and hydrogen.

■ **hydrocephalus** Commonly known as water on the brain, an excess of **cerebrospinal fluid** within the **ventricles** of the brain, eventually leading to malformation of brain tissue.

■ **hydrogen ion** H^+ Positively charged hydrogen atom; a proton. A characteristic of an **acid** is the production of hydrogen ions, which in aqueous solution are hydrated to hydroxonium ions, H_3O^+. Hydrogen ion concentration is a measure of acidity, usually expressed on the **pH** scale. A hydrogen ion concentration of 10^{-7} mol dm^{-3} (corresponding to pH 7) is neutral.

hydrological cycle Alternative name for **water cycle**.

■ **hydrolysis** Chemical decomposition of a substance by water, with a hydroxyl group (–OH) from the water taking part in the reaction; *e.g.* esters hydrolyse to form alcohols and acids.

■ **hydrophobia** Popular name for **rabies**, although in fact it refers to only one of the symptoms, the fear of drinking water.

■ **hydrophyte** Plant that lives in water. [2.5.a]

■ **hydrosol** Aqueous solution of a **colloid**.

hydrosphere Portion of the Earth's crust that consists of water: the oceans, seas, lakes, rivers, etc.

hydrostatic skeleton Supporting structure, consisting of fluid under pressure, possessed by some invertebrate animals (*e.g.* sponges).

■ **hydrotropism** Growth in plants in response to the presence of water; *e.g.* roots are hydrotropic.

■ **hygiene** Practice of contributing to good health through cleanliness, particularly of one's body and food.

■ **hypermetropia** Long-sightedness, a visual defect in which the eyeball is too short (front to back) so that light rays entering the **eye** from nearby objects would be brought to a focus at a point behind the retina. It can be corrected by spectacles or contact lenses made from converging (convex) lenses. Alternative name: hyperopia. *See also* **myopia**. [15.6.c]

■ **hypha** Microscopic hollow filament characteristic of **fungi**. Hyphae form a network called a mycelium. [2.3.b]

■ **hypogeal** Describing plant germination in which the **cotyledons** remain underground. *See also* **epigeal**. [3.4.d]

■ **hypophysis** Alternative name for **pituitary**.

■ **hypothalamus** Floor and sides of the vertebrate forebrain, which is concerned with physiological co-ordination of the body; *e.g.* regulation of body temperature, heart rate, breathing rate, blood pressure, sleep pattern as well as drinking, eating, water excretion and other metabolic functions.

hypothermia Condition in which body temperature is much lower than normal, as a result of which metabolic processes slow down. It may be induced deliberately to treat certain disorders (such as high fever), or may arise accidentally through exposure to freezing temperatures.

I

Ichneumonoidea Large family of parasitic wasps, often misleadingly referred to as ichneumon flies. A female ichneumonoid lays her eggs inside the body of a host (a spider or an insect) that is specific to her sub-family. When the eggs hatch, the larvae devour the host from within.

■ **id** Term used by some psychologists to denote the part of the human mind that is the source of primitive **instincts** and urges, and drives the **unconscious**. *See also* **ego**.

■ **identical twins** Two offspring that arise from a single fertilized ovum (**zygote**) by mitotic division of the zygote or during the early embryo stage. Alternative name: maternal twins. *See also* **twins**. [3.4.c]

■ **ileum** Last section of the small **intestine** in mammals, where both digestion and absorption take place. *See also* **duodenum**; **jejunum**. [3.4.a] [3.5.e] [3.8.a]

■ **imago** Sexually mature adult form of an insect that develops after previous stages of **metamorphosis**.

■ **imbricate** Describing any structure made up of overlapping parts, *e.g.* fish scales.

imido-urea Alternative name for **guanidine**.

■ **immune response** Response of vertebrates to invasion by a foreign substance (**antigen**). It involves the production of specific **antibody** molecules, which combine with the antigen to form an antigen-antibody complex. Antibodies may be present in body fluids or carried by **lymphocytes**. [3.8.b]

■ **immune system** Organs, substances and mechanisms involved in **immunity** in a particular organism. [3.8.b]

■ **immunity** Protection by an organism against infection. Defence may be divided into passive and active mechanisms. Passive processes prevent the entry of foreign invasion, *e.g.* skin, mucous membranes. Active mechanisms include **phagocytosis** by **leucocytes** and the **immune response** in animals. Plants can have immunity, *e.g.* by means of **phytoalexins**. [3.8.b]

■ **immunization** Stimulation of active artificial **immunity** by injection of small amounts of **antigen** (a **vaccine**). A specific **antibody** or immunoglobulin is formed in response and may persist in the body, preventing further infection by the same organism. [3.8.b]

■ **immunoglobulin** *See* **antibody**; **immunization**.

■ **immunology** Scientific study of **immunity**.

■ **immunosuppressive** Describing a drug that suppresses the **immune response**, given to recipients of transplanted organs to minimize that chance of rejection.

■ **implantation** Process in which a fertilized ovum (egg) or **embryo** becomes attached to the lining of the uterus (womb) of a mammal. It is the beginning of pregnancy. [3.4.c]

■ **imprinting** Behavioural attachment of an animal to a parent figure for protection. It occurs during the early stages of the life of many animals. [3.6.c]

■ **impulse** In biology, transmission of a message along a **nerve fibre**. The nerve impulse is an electrical phenomenon which results in depolarization of the nerve membrane. This **action potential** lasts for a millisecond before the **resting potential** is restored. *See also* **all-or-none response**. [3.7.c]

■ **inbreeding** Reproduction between closely related organisms of a species. *See also* **outbreeding**. [3.4.c]

■ **incisor** Chisel-shaped cutting tooth of mammals located at the front of the upper and lower jaws; human beings have eight incisors. They grow continually in rodents, which use them for gnawing. [3.4.b]

■ **incompatibility** *1.* Failure of a group of plants, algae or fungi to achieve fertilization over a period of time. *2.* Mismatching of biochemical components, *e.g.* between **blood groups**, or between a transplanted donor organ and its recipient, or between a **scion** and the plant on which it is grafted. With an organ transplant it can lead to immunological rejection (counteracted by **immunosuppressive drugs**). [4.4.a]

■ **incomplete dominance** Condition in which neither of a pair of **alleles** (genes) is **dominant**; their effects merge to produce an intermediate characteristic. [4.8.a]

■ **incubation** *1.* Period of time between laying and hatching in which an embryo (of a reptile or bird) develops in its egg. *2.* Period of time during which bacteria develop in a host's body before symptoms of an illness appear. [3.4.c]

■ **incus** One of the **ear ossicles**. Alternative name: anvil. [14.6.c]

■ **indehiscent** Describing a **fruit** that does not spontaneously open to release its seeds (*i.e.* the seed or seeds remain inside the fruit when it falls from the plant). [3.4.d]

■ **independent assortment** The second of **Mendel's laws**, which states that **genes** are transmitted independently from parents to offspring and assort freely. Thus there is an equal chance of any particular gene being transmitted to the **gametes**. It does not apply to genes that exhibit **linkage**. [4.8.a] [4.10.a]

■ **indicator** *1.* In biology, organism that survives only in certain environments; its presence gives information about the environment. *E.g.* the presence in water of certain bacteria that normally live in faeces indicates that the water is polluted with sewage. *2.* In chemistry, substance that changes colour to indicate end of a chemical reaction or the **pH** of a solution; *e.g.* litmus.

■ **infection** Illness that results from contracting a **pathogen** (*e.g.* bacterium, virus) from somebody already suffering from the disorder. *See* **carrier**; **vector**. [3.8.b]

■ **inflammation** Response in vertebrates to injury or local invasion by foreign bodies, resulting in heat, redness and swelling. It involves dilation of **blood** vessels, migration of **leucocytes** to the site of injury, and movement of fluid and **plasma proteins** into the inflamed tissue. [3.8.a]

■ **inflorescence** Any of the various types of arrangement of flowers on a plant's (single) main stem. [3.4.d]

■ **infrasound** Sound waves with a frequency below the threshold of human hearing, *i.e.* less than about 20 Hz. [14.6.b]

■ **inheritance** Process by which characteristics are passed on from parents to offspring, *i.e.* from one generation to another. [4.6.b]

inhibitor Substance that prevents an action from occurring; *e.g.* some germinating seeds produce an inhibitor in the soil that stops other seeds from germinating. [4.8.b]

■ **inner ear** Fluid-filled part of the **ear** that contains both the organs that convert sound to nerve impulses and the organs of balance. [3.4.a] [3.5.e] [3.8.a]

■ **inoculation** *See* **vaccine**. [14.6.b]

■ **insect** Arthropod animal of the class **Insecta**. [3.8.b]

■ **Insecta** Large class of arthropods. Most adult insects possess a three-part segmented body consisting of a **head, thorax** and **abdomen**, with an **exoskeleton** composed of **chitin**, one pair of **antennae**, and three pairs of walking legs; many also possess wings, typically two pairs. Respiration is by means of **tracheae**. Many insects have life-cycles that involve **metamorphosis**. [2.3.b] [2.4.a] [2.5.b]

■ **insecticide** Substance used to kill insects. There are two main types: those that are eaten by insects (after being applied to their food) or inhaled by insects, and those that kill by contact. Non-biodegradable insecticides (such as DDT) may persist for a long time and become concentrated in **food chains**, where they have a damaging ecological effect. [2.5.c] [2.9.b] [2.10.b]

■ **Insectivora** Order of insect-eating mammals. Insectivores include shrews, hedgehogs and moles. They resemble the ancestral mammals that co-existed with the dinosaurs. [2.5.d] [2.9.b]

■ **insectivore** Strictly an animal of the order **Insectivora**, although often extended to include any insect-eating animal (such as ant-eaters and some bats).

insemination Placing of **sperm** within the body of a female so that fertilization of an **ovum** (egg) is probable. [3.4.c]

■ **instar** The form assumed by an insect during any of the stages of its development prior to its final moult (*see* **metamorphosis**).

■ **instinct** General term for any animal behaviour, or fixed response to a particular stimulus, that has not been learned but inherited; it is therefore common to all the members of a species. Instinct can be modified by learning; *e.g.* a bird instinctively knows how to sing but may have to learn a particular song. [3.7.c]

■ **insulin Hormone** secreted in vertebrates that controls the level of **glucose** in the blood. It is produced by groups of cells, called islets of Langerhans, in the **pancreas**. A deficiency of insulin results in the disorder diabetes mellitus, whose symptoms include excessive thirst and high levels of glucose in the blood and urine. [3.4.a] [3.5.c] [3.7.b] [3.8.a]

■ **integument** *1.* Animal's outer protective covering; *e.g.* cuticle, skin. *2.* Part of the outer coating of a seed (*see* **testa**). [3.4.d]

■ **intercellular** Between cells; *e.g.* intercellular fluid surrounds cells, maintaining a constant internal environment. *See also* **intracellular**.

■ **intercostal muscle** Any of the muscles between a mammal's ribs, which are important in breathing movements. [3.5.e] [3.6.a]

■ **interferon** Substance that is produced by animal cells, which prevents the multiplication of **viruses**. It is a protein, and its action is not specific to any particular group of viruses. There is some debate as to whether it may be useful in the treatment of certain forms of cancer. [3.8.b]

■ **intermediate host** Organism that acts as host to a maturing **parasite** (*e.g.* to the larval stage of an insect), but which is not a host to the mature parasite. [2.5.a]

■ **interphase** State of cells when not undergoing division. Preparation for division (**mitosis**) is carried out during this phase, including replication of **DNA** and cell constituents. [4.7.a]

■ **interstitial cells** In mammals, cells present in the male and female **gonads**. In males they are found between the **testis** tubules and in females in the **ovarian follicle**. When stimulated by **luteinizing hormone**, they produce **androgens** in males and **oestrogen** in females.

■ **intestine** In vertebrates, the part of the gut between the **stomach** and **anus**. It usually consists of two sections: the small and the large intestine. The small intestine is concerned with the further digestion of food and the absorption into the bloodstream of already digested **amino acids** and **monosaccharides**. The large intestine is mainly concerned with the absorption of water from semi-solid indigestible remains, which form the **faeces**. [3.4.a] [3.5.e] [3.7.b] [3.8.a]

■ **intracellular** Occurring within the boundary of a cell or cells. *See also* **intercellular**.

■ **inversion** In organic chemistry, splitting of **dextrorotatory** higher sugars (*e.g.* sucrose) into equivalent amounts of **laevorotatory** lower sugars (*e.g.* fructose and glucose).

■ **invertase** Plant **enzyme** that brings about the hydrolysis of **sucrose** (cane-sugar) to form **invert sugar**.

■ **invertebrate** Animal that does not possess a backbone. *See also* **vertebrate**. [2.3.b] [2.4.a] [2.5.b]

■ **invert sugar** Natural **disaccharide** sugar that consists of a mixture of **glucose** and **fructose**, found in many fruits. It is also formed by the hydrolysis of **sucrose** (cane-sugar), which can be brought about by the enzyme **invertase**.

■ **in vitro** Literally 'in glass'. It describes techniques or processes that are carried on outside a living organism, *e.g.* in a test-tube.

■ **in vivo** Literally 'in life'. It describes biological experiments or processes that are carried on inside living organisms.

■ **involuntary muscle** Muscle not under conscious control, located in internal organs and tissues, *e.g.* in the alimentary canal and blood vessels. Alternative name: smooth muscle. *See also* **voluntary muscle**. [3.5.e] [3.7.b]

■ **ion** Atom or molecule that has a positive or negative electric charge because of the loss or gain of one or more electrons. Many inorganic compounds dissociate into free ions when they dissolve in water; *e.g.* sodium chloride (NaCl) forms sodium ions (Na^+) and chloride ions (Cl^-) in solution.

■ **iris** Pigmented part of the human **eye** that controls the amount of light entering the eye. *See also* **iris diaphragm**. [15.6.b]

■ **irritability** Ability to respond to a stimulus, evident in all living material. [3.7.c]

■ **islets of Langerhans** In vertebrates, group of specialized secretory cells in the **pancreas**. They control the level of **glucose** in the **blood** by secreting **insulin** and **glucagon**. These hormones in turn either stimulate the liver to convert glucose to glycogen (function of insulin) or cause glycogen to be broken down to release glucose (function of glucagon). [3.7.b] [3.8.a] [3.9.b]

■ **isogamy** Sexual fusion of **gametes** that are similar in structure and size. It occurs in some protozoa, fungi and algae. *See also* **anisogamy**.

■ **isomorphism** In plants, existence of morphologically identical male and female individuals or generations.

isopod Crustacean of the order **Isopoda**.

Isopoda Order of small **crustaceans**. Most of the 4,000 species of isopods are marine, but some, notably woodlice, have adapted to life on land.

Isoptera Order of social insects related to cockroaches, which includes termites.

■ **isotonic** Having the same **osmotic pressure** as blood, or the same as the cell sap in a particular plant. [3.5.c] [3.8.b]

J

■ **jaws** *1.* In vertebrates, bony structure enclosing the mouth, often furnished with teeth for grasping prey and/or chewing food, consisting of an upper **maxilla** and lower **mandible**. *2.* In invertebrates, grasping structure surrounding the mouth. [3.4.b] [3.5.e] [3.7.b]

■ **jejunum** Part of the **intestine** of mammals, located between the **duodenum** and the **ileum**, the main function of which is absorption. [3.7.b]

■ **joint** In anatomy, point of articulation of limbs or **bones**. The bones are connected to each other by connective tissue **ligaments**, and are well lubricated. Common types of joints include ball-and-socket joints (*e.g.* the hip and shoulder joints), hinge joints (*e.g.* the elbow and knee joints) and sliding joints (*e.g.* between vertebrae). [3.3.a] [3.5.e]

■ **joule** (J) SI unit of **work** and **energy**. 4.2 J = 1 calorie; 4.2 kilojoules = 1 kilocalorie, or 1 Calorie (with a capital C). [3.6.a] [3.7.b]

■ **jugular vein** In vertebrates, main **vein** carrying blood from the head (including the brain) to the heart. [3.5.d] [3.5.e]

■ **jungle** Form of dense tropical vegetation, a tangle of creepers (lianas), bamboo, shrubs and palms. It occurs in areas of rainforest that have not yet reached the status of a **climax community**, typically on the site of a former clearing. [2.2.a] [2.5.a]

■ **juvenile hormone** In insects, a **hormone** that is secreted at certain stages of development. The hormone inhibits **metamorphosis** to the adult form of the insect, but promotes larval growth and development. *See also* **neoteny**.

K

karyokinesis *See* **mitosis**.

karyotype Visible trait on the **chromosome** of a typical **cell**. It provides information about the species or strain, because a karyotype is characteristic to the cell of a particular organism.

■ **keratin** Tough, non-living outermost layer of skin which forms a protective covering in vertebrates. Formed from **epidermis**, it may be modified to make *e.g.* hair, feathers, horns or claws.

■ **keratinization** Replacement of the **cytoplasm** of cells in the **epidermis** by **keratin**, thus resulting in hardening of skin. Alternative name: cornification.

■ **ketonuria** Presence in the urine of compounds containing **ketones**, usually associated with the disorder diabetes mellitus. A newborn baby's urine is tested for ketonuria soon after birth. [3.8.a] [3.8.b]

key Set of paired descriptions used to identify an organism. [2.5.b]

■ **kidney** In vertebrates, one of a pair of excretory organs that filter waste products (particularly nitrogenous waste) from the blood and concentrate them in **urine**. Kidneys also have an important function in the regulation of the balance of water and salts in the body. The main processes of the kidney occur in a large number of tubular structures called nephrons. Water and waste products pass from the kidneys to the bladder via the ureters. [3.4.a] [3.8.a]

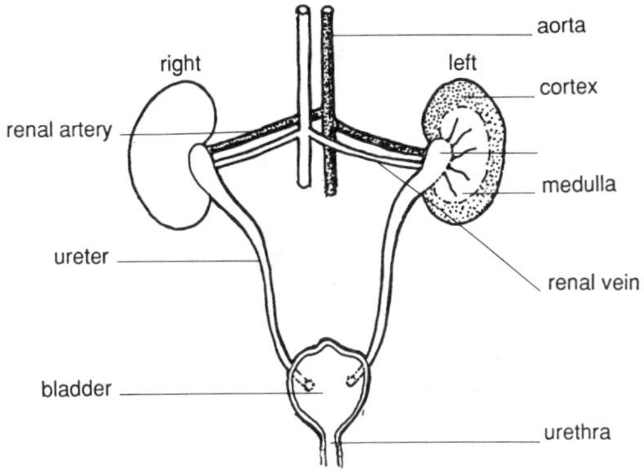

The human kidneys

■ **kilocalorie** (C) Unit of energy equal to 1,000 **calories**. 1 C = 4.2 kilojoules. Alternative name: Calorie (capital C).

■ **kilojoule** (kJ) Unit of energy equal to 1,000 **joules**.

■ **kinaesthesis** Process by which sensory cells in muscles and organs relay information concerning the relative position of the limbs and the general orientation of an organism in space; a type of biological **feedback**.

■ **kinase Enzyme** that causes **phosphorylation** by ATP.

■ **kinesis** In zoology, simplest kind of orientation behaviour that occurs in response to a stimulus (*e.g.* the concentration of a nutrient or an irritant). The speed of an animal's random motion increases until the stimulus reduces. *See also* **taxis**.

kinetochore Point of attachment of the **spindle** in a **cell**.

■ **king crab** Large member of a class of marine **arthropods**, most of which are extinct, that are related to the scorpions. Alternative name: horseshoe crab.

■ **kingdom** Highest rank in the **classification** of living organisms, which encompasses phyla (for animals) and divisions (for plants). Criteria determining members of a kingdom are broad, and consequently members are very diverse. Traditionally, there were two kingdoms: the plants and the animals. However, **fungi**, **protists** and **prokaryotes** are now often placed in kingdoms of their own. [2.3.b] [2.4.a] [2.5.b]

■ **kinin** *1.* In plants, growth substance that stimulates cell division; if naturally occurring it is termed a cytokinin; *e.g.* zeatin. *2.* In animals, **peptide** found in blood, possibly secreted as a response to inflammation caused by stings and venoms; it causes contraction of smooth muscles and dilation of blood vessels.

klino-taxis Movement of an animal in respose to light.

■ **Krebs cycle** Energy-generating series of biochemical reactions that occur in living cells. It forms the second stage of **aerobic respiration**, in which **pyruvate** or **lactic acid** produced by **glycolysis** is oxidized to carbon dioxide and water, thus producing a large amount of energy in the form of **ATP** molecules. The Krebs cycle takes place in **mitochondria**. It was named after the German-born British biochemist Hans Krebs (1900–82). Alternative names: citric acid cycle, tricarboxylic acid cycle. [3.6.a] [3.6.b]

k-selection Survival strategy based on maximum competitiveness that is one of the processes in **natural selection**. K-selection species are characterized by low birth

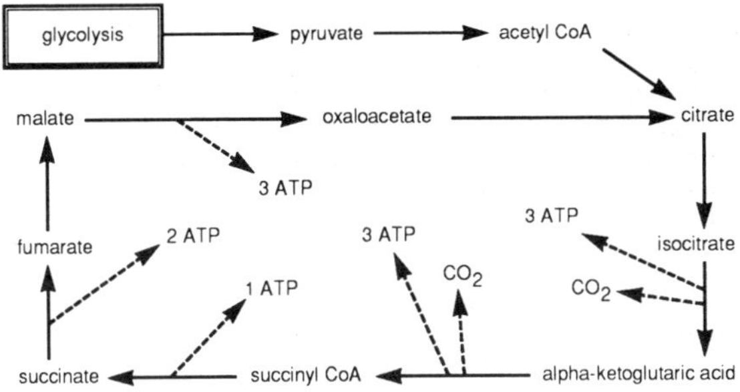

The Krebs cycle

rates, prolonged development of young, and high survival rates for offspring. *See also:* **r-selection**.

L

■ **labelling** Technique in which an atom (often in a molecule of a compound) is replaced by its radioisotope (termed a tracer) as a means of locating the compound or following its progress; *e.g.* in a plant or animal, or in a chemical reaction.

■ **labiate** *1.* Having lips or structures resembling lips. *2.* Member of a family of herbs and small shrubs (Labiatae), some of which are cultivated for the aromatic oils, *e.g.* menthol, produced in their flowers. [2.5.b]

■ **labium** *1.* In insects, the lower lip, used for manipulating food during feeding. *2.* In mammals, one of the folds of flesh at the entrance to the **vagina**. [3.4.c]

■ **labrum** *1.* In crustaceans and insects, a ridge of **cuticle** that forms the upper lip. *2.* In gastropod molluscs, the outer margin of the shell.

■ **labyrinth** Part of the inner **ear** of vertebrates concerned with balance, consisting of three semi-circular canals (each in a different plane) containing fluid. A chalky body (**otolith**) in each canal moves against sensitive receptors, so detecting movements of the head. [3.4.a] [3.5.e] [3.8.a] [14.6.b]

■ **lacrimal gland** Gland that produces tears. Fluid is continuously secreted to protect and moisten the **cornea**; it also contains the bactericidal enzyme **lysozyme**. Alternative name: lachrimal gland.

β-lactam Member of a group of **antibiotics** that include the penicillins.

lactate *1.* Salt or ester of **lactic acid** (2-hydroxypropanoic acid). *2.* To produce milk.

■ **lactation** Milk-production; it occurs in mammals for feeding the young. [3.4.c] [3.5.b] [3.6.c]

■ **lacteal** **Lymph** vessel of the **villi** in the **intestine** of vertebrates. Fat passes into the lacteals as an emulsion of globules to be circulated in the **lymphatic system**. [3.7.b]

■ **lactic acid** $CH_3CH(OH)COOH$ Colourless liquid organic acid. It is produced in animals when **anaerobic respiration** takes place in muscles because of an insufficient oxygen supply during vigorous activity. Lactic acid is also formed by the action of bacteria on the lactose in milk when it turns sour. [3.6.a]

lactoflavin Alternative name for **riboflavin**.

lactose $C_{12}H_{22}O_{11}$ White crystalline **disaccharide** sugar that occurs in milk, formed from the union of **glucose** and **galactose**. It is a reducing sugar. Alternative name: milk-sugar.

laevorotatory Describing a compound with **optical activity** that causes the plane of polarized light to rotate in an anti-clockwise direction. Indicated by the prefix (–)- or *l*-.

laevulose Alternative name for **fructose**.

■ **Lamarckism** Theory proposed by the French naturalist Jean-Baptiste Lamarck (1744–1829) that evolutionary change could be achieved by the transmission of acquired characteristics from parents to offspring. The theory was superseded by **Darwinism**, and there has yet to be any firm evidence that the inheritance of acquired characteristics ever occurs at all. *See also* **Lysenkoism**. [4.9.a]

■ **lamella** Thin plate-like structure, *e.g.* the membrane in a **chloroplast** that forms folded structures containing chlorophyll. [3.6.b] [3.7.a] [3.8.a]

lamellibranch Alternative name for a bivalve **mollusc**.

■ **lamprey** Jawless parasitic eel-like fish of the class Agnatha.

■ **lanolin** Yellowish sticky substance obtained from the grease that occurs naturally in wool. It is used in cosmetics, as an ointment and in treating leather. Alternative names: lanoline, wool fat.

■ **large intestine** *See* **colon**; **intestine**. [3.4.a] [3.5.c] [3.5.e] [3.7.a] [3.8.a]

■ **larva** Juvenile form of some animals which, while radically different from the adult in form, is capable of independent existence; *e.g.* a caterpillar, grub, maggot or tadpole. The larva changes to the adult by **metamorphosis**, *e.g.* in some insects and amphibians. *See also* **imago**; **pupa**.

■ **larynx** Region of **trachea** (windpipe) that usually houses the **vocal cords** (composed of membrane folds that vibrate to produce sounds). [3.5.e]

■ **lateral line** Line of receptor cells, sensitive to water pressure and vibration, that is found along the sides of the body and head of all fish and some amphibians. [3.5.e]

■ **latex** Milky fluid produced in some plants after damage, containing sugars, proteins and alkaloids. It is used in manufacture, *e.g.* of rubber. A suspension of synthetic rubber is also called latex.

LD – 50 (lethal dose 50) Toxicity test that estimates the quantity of a substance needed to cause death in 50% of the organisms tested.

■ **leaf** Part of a plant, usually flat and green, that grows from the stem. In green plants leaves are the sites of **photosynthesis** and **transpiration**. [3.6.b] [3.7.a] [3.9.a]

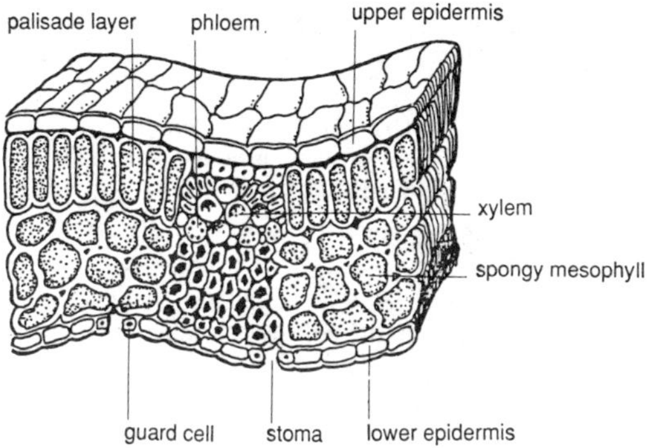

Section of a leaf

■ **leaf litter** Layer of decaying plant material on the surface of the soil. *See also* **humus**.

lecithin Type of **phospholipid**, a **glyceride** in which one organic acid residue is replaced by a group containing **phosphoric acid** and the base **choline**. It is a major component of a cell **membrane**. Alternative name: phosphatidyl choline.

■ **leguminous** Describing any plant of the pea family (legumes), *e.g.* beans, clover, lucerne (alfalfa), peas. [2.7.a]

■ **lenticel** Raised pore on the surface of a woody stem that allows entry and exit of gases. [3.6.a] [3.7.a]

■ **Lepidoptera** Large order of insects, consisting of moths and butterflies, characterized by wings made up of overlapping scales. [2.3.b] [2.4.a] [2.5.b]

■ **leptotene** Stage of **prophase** in the first cell division in **meiosis**. At this stage, **chromosomes** can be seen to carry chromomeres. [4.6.a]

leucine $(CH_3)_2CHCH_2CH(NH_2)COOH$ Colourless crystalline **amino acid**; a constituent of many **proteins**. Alternative name: 2-amino-4-methylpentanoic acid.

■ **leuco-** Prefix denoting white; *e.g.* leucocyte. (The prefix is spelled *leuk-* in *leukaemia*.)

■ **leucocyte** White blood cell. Leucocytes are generally amoeboid, and play a major role in defending the body against disease.

■ **leucoplast** Colourless **plastid**, often containing reserves of starch.

■ **leukaemia** Disorder characterized by overproduction of white blood cells, often called blood cancer. There are many forms, some of which can now be successfully treated. [3.8.a]

■ **lichen** Organism formed from a symbiotic relationship (*see* **symbiosis**) between a **fungus** and an **alga**, classified in the plant division Lichenes. [2.3.a] [2.5.a]

Lichenes Division of the plant kingdom that contains **lichens**.

■ **life cycle** Progressive sequence of changes that an organism undergoes from fertilization to death. In the course of the cycle a new generation is usually produced. Reproduction may be sexual or asexual; both **meiosis** and **mitosis** may occur. *See also* **alternation of generations**. [2.5.a] [2.5.b] [3.4.c]

■ **ligament** Tough elastic connective tissue that connects bones together at joints. *See also* **tendon**. [3.3.a] [3.5.e]

ligand In biochemistry, any molecule that interacts with or binds to a receptor that has an affinity for it.

ligase Enzyme that repairs damage to the strands that make up **DNA**, widely used in **recombination** techniques to seal the joins between DNA sequences.

■ **lignin** Complex **polymer** found in many plant **cell walls**, that glues together fibres of **cellulose** and provides additional support for the cell wall. [3.7.a]

Liliopsida Class of **angiosperm** plants that comprises the **monocotyledons**.

line transect In an ecological survey, a method of systematically sampling along a line or narrow band.

■ **linkage** Occurrence of two **genes** on the same **chromosome**. Genes that are close together are likely to be inherited together; genes that are farther apart may become separated during **crossing over**. [4.9.d]

■ **Linnaean system** System that classifies and names all organisms according to scientific principles. Each species has two names; the first indicates the organism's general type (genus), the second gives the unique species, e.g. *Canis* (dog) *domesticus* (household). The system was called after the Swedish botanist Carl Linné (Linnaeus) (1707–78). Alternative name: binomial classification. [1.7.f] [2.3.b] [2.4.a] [2.5.b]

■ **lipase** In vertebrates, an **enzyme** in intestinal juice and pancreatic juice that catalyses the **hydrolysis** of **fats** to **glycerol** and **fatty acids**. [3.7.b]

■ **lipid** Member of a group of naturally occurring fatty or oily compounds that share the property of being soluble in organic solvents, but sparingly soluble in water. Also, all lipids yield **monocarboxylic acids** on **hydrolysis**.

lipochrome Yellow pigment in butterfat.

lipolyte **Lipid**-containing **cell**. Alternative name: fat cell.

liposome Droplet of **fat** in the **cytoplasm** of a cell, particularly that of an **egg**.

■ **liver** In vertebrates, a large organ in the abdomen, the main function of which is to regulate the chemical composition of the blood by removing surplus **carbohydrates** and **amino acids**, converting the former into **glycogen** for storage and the latter into **urea** for excretion. Other functions include storage of **iron**, metabolism and storage of **fats**, secretion of **bile** and production of the blood-clotting factors prothrombin and **fibrinogen**. [3.4.a] [3.5.e] [3.7.b] [3.8.a]

■ **locomotion** Movement from place to place, one of the distinguishing characteristics of animals (as opposed to plants).

■ **locus** In biology, position of a **gene** on a **chromosome**. [4.6.a]

■ **long sight** Alternative name for **hypermetropia**. [15.6.c]

■ **LSD** Abbreviation of **lysergic acid diethylamide**. [3.4.b]

luciferase **Enzyme** that initiates the **oxidation** of **luciferin**.

■ **luciferin** Substance that occurs in the light-producing organ of some animals, *e.g.* firefly. When oxidized (through the action of the enzyme luciferase) it produces **bioluminescence**.

■ **lumen** In biology, the space enclosed by a duct, vessel or tubular organ.

■ **lung** In air-breathing vertebrates, paired or single respiratory organ located in the **thorax**. Its surface contains a large area of thinly folded, moist **epithelium** membrane so that it occupies little volume. This membrane is richly supplied by

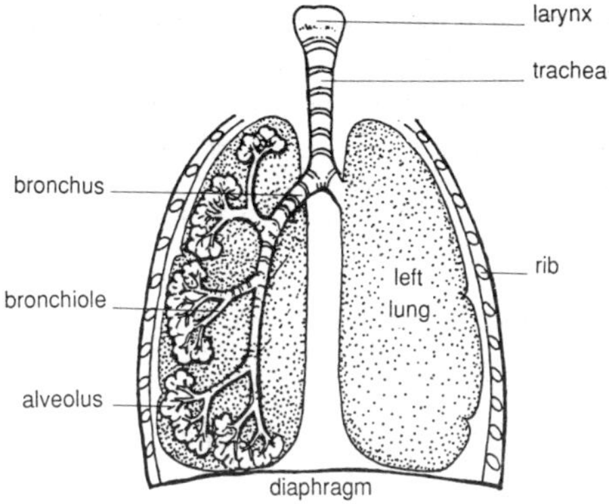

Human lungs

blood **capillaries** which allow for efficient and easy gaseous exchange. Air enters and leaves lungs through the **bronchus**, which in mammals branches into **bronchioles** ending in clusters of **alveoli**, where the main gaseous exchange takes place. [3.4.a] [3.5.e] [3.6.a] [3.8.a]

lutein Alternative name for **xanthophyll**.

■ **luteinizing hormone** (LH) **Glycoprotein hormone** that is secreted by the **pituitary** under regulation by the **hypothalamus**. In female mammals, it stimulates the formation of the **corpus luteum** and **Graafian follicle**, which in turn stimulates **oestrogen** production. LH is also essential for **ovulation**. In males, it stimulates interstitial cells of the **testis** to produce **androgens**. [3.4.c]

luteotrophin Alternative name for **prolactin**.

Lycopsida One of the most primitive classes of vascular plants, the club mosses.

■ **lymph** Colourless alkaline tissue fluid that drains into the lymphatic system from intercellular spaces. It is similar in salt concentration to **plasma**, but possesses a lower **protein** concentration. It contains cells that are chiefly **lymphocytes**. [3.5.c] [3.8.b]

■ **lymphatic system** System of vessels and nodes (glands) that circulates **lymph** throughout the body, transporting the products from the digestion of fats and producing **antibodies** and **leucocytes**. [3.5.c] [3.8.b]

■ **lymph node** Flat oval structure made of lymphoid tissue that lies in the lymphatic vessels and occurs in clusters in the neck, armpit or groin. Its main function is the manufacture of **antibodies** and **leucocytes**. Lymph nodes also act as a defence barrier against the spread of infection by filtering out foreign bodies and bacteria, thus preventing their entry into the bloodstream. Alternative name: lymph gland. [3.5.c] [3.8.b]

■ **lymphocyte** Type of agranular **leucocyte** with a large nucleus, rich in **DNA**. Lymphocytes form 25% of all leucocytes and produce **antibodies**, which are important in defence against disease. They are produced in **myeloid tissue** of red bone marrow, **spleen**, **tonsils**, **lymph nodes** and **thymus**. [3.5.c] [3.8.b]

■ **lymphoid tissue** Tissue found in dense aggregations in **lymph nodes**, **tonsils**, **thymus** and **spleen**. It produces **lymphocytes** and macrophagocytic cells, which ingest **bacteria** and other foreign bodies. Alternative name: lymphatic tissue. [3.5.c] [3.8.b]

lyophilic Possessing an affinity for liquids.

lyophobic Liquid-repellent, having no attraction for liquids.

Lysenkoism Doctrine adopted in the Soviet Union from the 1940s to the 1960s that held all variation in organisms to be the direct result of environmental influence, denied the existence of genes and revived the tenets of **Lamarckism**. The doctrine was adopted for purely political reasons and named after its main advocate, Trofim Lysenko (1898–1976).

■ **lysergic acid diethylamide** (LSD) Synthetic substance, similar to some fungus **alkaloids**, which provokes hallucinations and extreme mental disturbance if taken, even in extremely small quantities.

lysine $H_2N(CH_2)_4CH(NH_2)COOH$ **Essential amino acid** that occurs in **proteins** and is responsible for their base-neutralizing powers because of its two $-NH_2$ groups. Alternative name: diaminocaproic acid.

■ **lysis** Degeneration and subsequent breakdown of a cell. Under normal conditions, cell lysis is carried out by **phagocytes**, which also lyse invading cells. Under rare conditions, lysis may occur from within a cell.

■ **lysosome** Membrane-bound organelle of cells that contains a range of digestive **enzymes**, such as **proteases**, phosphatases, **lipases** and **nucleases**. The functions of lysosomes include contributing enzymes to white **blood cells** during **phagocytosis** and the destruction of cells and tissue during normal development, *e.g.* the loss of a tadpole's tail. Lysosomes may be produced directly from the **endoplasmic reticulum** or by budding of the **Golgi apparatus**.

lysozyme **Enzyme** in saliva, egg white, tears and mucus. It catalyses the destruction of bacterial cell walls by **hydrolysis** of their mucopeptides, and thus has a bactericidal effect.

M

■ **macromolecule** Very large **molecule** containing hundreds or thousands of **atoms**; *e.g.* natural **polymers** such as cellulose, rubber and starch, and synthetic ones, including **plastics**.

macronutrient Food substance needed in fairly large amounts by living organisms, which may be an inorganic element (*e.g.* phosphorus or potassium in plants) or an organic compound (*e.g.* amino acids and carbohydrates in animals). *See also* **trace element; vitamin**.

macrophage Large **phagocyte** found in vertebrate tissue, particularly in liver, spleen and lymph glands. It removes foreign particles at the site of infection.

■ **macula** Any small blemish, coloured patch, growth or shallow depression that is visible on plant or animal tissue, especially on external surfaces. *E.g.* the macula lutea is the yellow spot (**fovea**) at the centre of the retina.

Magnoliophyta Subdivision of the plant division **Spermatophyta**, consisting of the **angiosperms**.

Magnoliopsida Class of **angiosperm** plants that comprises the **dicotyledons**.

Malacostraca Subclass of crustaceans that includes lobsters, crabs and shrimps.

■ **malignant** Describing a (medical) condition that tends to spread or get worse; opposite of benign.

■ **malleus** One of the **ear** ossicles. [14.6.b]

Malpighian body Part of the mammalian **kidney**, encompassing the **Bowman's capsule** and **glomerulus**. Its function is to filter blood. It was named after the Italian biologist Marcello Malpighi (1628–1694).

■ **maltose** $C_{12}H_{22}O_{11}$ Common **disaccharide** sugar, composed of two molecules of **glucose**. It is found in **starch** and **glycogen**, and used in the food and brewing industries.

■ **mammal** Member of the animal class **Mammalia**. [2.3.b] [2.4.a] [2.5.b]

■ **Mammalia** Class of tetrapod vertebrates which includes man. General characteristics of mammals include warm-bloodedness, a four-chambered heart, the possession of hair, the possession of middle-ear ossicles, a diaphragm used in respiration, giving birth to live young (nourished in the womb via a placenta), and considerable parental care including feeding of young with secreted milk. [2.3.b] [2.4.a] [2.5.b]

■ **mammary gland** Gland for milk-production (lactation) in female mammals. [2.5.b] [3.4.c]

mammotrophin Alternative name of **prolactin**.

■ **mandible** *1.* Mouth-part in animals such as crustaceans, insects and cephalopods, used for cutting food. *2.* Lower **jaw** of a vertebrate. [3.5.e]

mannitol $HO.CH_2(CHOH)_4CH_2OH$ Soluble hexahydric alcohol that occurs in many plants, and is used in medicine as a diuretic sugar substitute.

■ **mantid** Winged insect that is characterized by a narrow, elongated body and powerful forelegs which are used for seizing insect prey. There are about 1,800 species, found throughout the warmer regions of the world, of which the praying mantis is the best known. [2.3.b] [2.4.a] [2.5.b] [3.7.b]

■ **mantle** *1.* Body part from which the shell of a mollusc or brachiopod develops. *2.* Alternative name for the **carapace** of an adult barnacle.

■ **mare** Adult female horse.

■ **marsupial** *1.* Member of the order **Marsupiala**. *2.* Animal that has a body part that bears a structural or functional resemblance to the pouch possessed by true female marsupials, *e.g.* marsupial frog. [2.3.b] [2.4.a] [2.5.b]

■ **Marsupiala** Order of mammals in the subclass Metatheria, now restricted to Australasia and South America, characterized by the development of the **embryo** after birth in an external pouch (marsupium) that surrounds the **mammary glands**. In Australasia, marsupials have evolved into a wide range of types – kangaroo, koala, Tasmanian wolf, marsupial rat, marsupial mouse, etc. In America, opossums are the only surviving members of the order. [2.3.b] [2.4.a] [2.5.b]

■ **maternal twins** *See* **twins**.

■ **matrix** In biology, extracellular substance that embeds and connects cells, *e.g.* connective tissue.

■ **maxilla** *1.* Mouthpart of an insect used in feeding. *2.* Upper **jaw** of a vertebrate. [3.7.b]

meatus Duct or channel between body parts, *e.g.* the auditory canal, which leads into the ear.

median lethal dose *See* **LD-50**.

■ **medicine** Science that studies diseases and other bodily disorders, how they are caused and how they are treated. In everyday language, a medicine is any **drug** or other remedy used to treat a disorder.

■ **medulla** Central part of an organ or tissue (*e.g.* adrenal medulla). *See also* **cortex**. [3.4.a] [3.5.e] [3.7.c] [3.8.a]

■ **medulla oblongata** Posterior part of the **brain** of vertebrates, which contains centres that control the frequency of respiration and heartbeat, and is concerned with the co-ordination of nerve impulses from hearing, touch and taste receptors.

■ **medullary ray** Strip of **parenchyma** cells that runs through tissues of plants, used for the storage and transport of food. It may be primary (running from the centre to the **cortex**), or secondary (formed by secondary thickening from **cambium**).

medullated nerve fibre *See* **myelin; nerve**.

■ **medusa Gamete**-producing form of a **coelenterate**, *e.g.* jellyfish, which thereby reproduces sexually. Other forms in the life-cycle reproduce asexually. *See also* **alternation of generations**.

■ **meiosis** Type of cell division in which the number of **chromosomes** in the daughter cells is halved; thus they are in the **haploid** state. Two successive divisions occur in the process, giving four daughter cells. The first division takes place in four stages: **prophase**, **metaphase**, **anaphase** and **telophase**. The second division has three stages: metaphase, anaphase and telophase. In animals meiosis occurs in the formation of **gametes**, *e.g.* eggs and sperm. Alternative name: reduction division. *See also* **mitosis**. [4.7.a]

■ **meiotic** Describing or referring to **meiosis**.

■ **melanin** Dark brown or black pigment that gives colour to hair and skin. It is formed in melanocytes by the oxidation of **tyrosine**, induced by the action of sunlight. [2.6.a] [3.8.a]

■ **melanism** Occurrence in animal populations of dark-coloured individuals having an excess of **melanin** in their tissues.

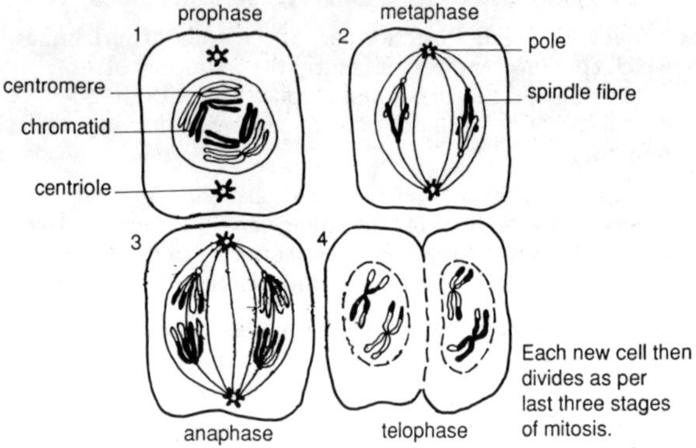

Stages of meiosis

melanocyte Animal cell that produces **melanin**.

■ **membrane** Thin film of tissue. In living organisms, membranes form a dynamic interface, *e.g.* the outer membrane between a cell and its surroundings (*see* **plasma membrane**). The structure may be selective in allowing the passage of certain molecules through (*e.g.* a semipermeable membrane) or specific (*e.g.* in active transport). Membranes are widely distributed and very important in all organisms.

membranous labyrinth *See* **ear**.

■ **Mendelism** Study of inheritance, and therefore genetics, as a result of work carried out by the Austrian monk Gregor Mendel (1822–84). *See also* **Mendel's laws**. [4.6.b] [4.8.b]

■ **Mendel's laws** Conclusions drawn from work on inheritance carried out by Gregor Mendel in breeding experiments. The first is the law of segregation: an inherited characteristic is controlled by a pair of factors (**alleles**), which separate and become incorporated into different **gametes**. The second is the law of independent assortment: the separated factors are independent of each other when gametes form. [4.8.a] [4.10.a]

■ **meninges** System of **membranes** that envelop the brain. The innermost is the pia mater, the outermost the dura mater. [3.7.c]

■ **menopause** Natural cessation of menstruation (*see* **menstrual cycle**). [3.4.c]

■ **menstrual cycle Hormone**-regulated cycle of female reproductive behaviour during which **ovulation** occurs; it occurs in some primates, including human beings. The end of the cycle is marked by a monthly shedding of the endometrium (lining of the womb) accompanied by a discharge of blood from the vagina; this is known as menstruation. The cycle begins with the menarche at the beginning of puberty and ends at the menopause. Alternative name: sexual cycle. *See also* **oestrous cycle**. [3.4.c] [3.6.f]

■ **menstruation** *See* **menstrual cycle**.

■ **meristem** Region of active cell division and differentiation in plants. The principal meristems in flowering plants occur at the tips of stems and roots.

mescaline Powerful drug derived from the Mexican mescal cactus (peyote), which has a similar effect to **lysergic acid diethylamide** (LSD), and causes hallucinations and mental disturbance.

mesencephalon Alternative name for **mid-brain**.

■ **mesentery** Vertical fold of tissue on the inner surface of the body wall of animals, which supports internal organs or associated structures.

■ **mesoderm** Tissue in an animal embryo that develops into tissues between the gut and **ectoderm**.

■ **mesoglea** Layer of unstructured material that occurs between the **ectoderm** and **endoderm** of **coelenterates** (*e.g.* jellyfish).

■ **mesophyll** Tissue that forms the middle part of a leaf.

■ **mesophyte** Plant that grows in an environment with an average water supply. *See also* **hydrophyte; xerophyte**. [2.6.a]

■ **messenger RNA** (mRNA) Ribonucleic acid that conveys instructions from **DNA** by copying the code of DNA in the cell **nucleus** and passing it out to the cytoplasm. It is translated into a **polypeptide** chain formed from **amino acids** which join in a sequence according to the instructions in the messenger RNA. *See also* **transcription**.

■ **metabolism** Biochemical reactions that occur in cells and are a characteristic of all living organisms. Metabolic reactions are initiated by **enzymes** and liberate energy in a usable form. Organic compounds may be broken down to simple constituents (catabolism) and used for other processes. Simple compounds may be built up to more complex ones (anabolism). [3.5.e] [3.8.a] [3.10.a]

■ **metabolite** Molecule participating in **metabolism**, which may be synthesized in an organism or taken in as food. Autotrophic organisms need only to take in inorganic metabolites; heterotrophs also need organic metabolites.

metameric segmentation Alternative name: metamerism. *See* **segmentation**.

■ **metamorphosis** Stage in the life cycle of some animals, including marine invertebrates, arthropods and amphibians, in which by hormonal control the **larva** undergoes transformation to the adult form.

■ **metaphase** Second stage of **mitosis** and **meiosis**, in which **chromosomes** are lined up along the equator of the nuclear **spindle**. Alternative name: aster phase. [4.7.a]

Metaphyta The plant kingdom. Alternative name: Plantae.

metaplasia Transformation of one normal tissue type into another as a response to a disease or abnormal condition.

metastasis Process by which disease-bearing cells are transferred from one part of the body to another via the **lymph** and **blood vessels**; the term is usually applied to the spread of **cancers**. The term also applies to the newly diseased area arising from the process.

■ **metatarsal** One of the rod-shaped bones that forms the lower hind limb or part of the hind foot in four-legged animals and the arch of the foot in human beings. [3.3.a]

Metatheria Subclass of mammals that contains the **marsupials**.

■ **Metazoa** Subkingdom consisting of multicellular animals with a body having two or more tissue layers and a co-ordinating nervous system. [2.3.b] [2.4.a] [2.5.b]

metazoan Animal that is a member of the **Metazoa**.

methionine $CH_3S(CH_2)_2CH(NH_2)COOH$ Sulphur-containing **amino acid**; a constituent of many **proteins**. Alternative name: 2-amino-4-methylthiobutanoic acid.

■ **microbe** Imprecise term for any **micro-organism**, particularly a disease-causing **bacterium**. [2.4.b] [2.7.a] [2.8.a]

■ **microbiology** Biological study of **micro-organisms.**

■ **micronutrient** General term for any of the **trace elements** or **vitamins.**

■ **micro-organism** Organism that may be seen only with the aid of a **microscope**. Micro-organisms include microscopic **fungi** and **algae, bacteria, viruses** and single-celled animals (*e.g.* protozoans). [2.7.a]

■ **micropyle** Small opening or pore. *1.* Pore in a flower's **ovule** through which the **pollen tube** enters (to bring about pollination). *2.* Pore in a flower's seed through which water enters (to initiate germination). *3.* Pore in an insect's egg through which sperm enter (to bring about fertilization). [3.4.d]

■ **microtome** Instrument for cutting thin slices (of the order of a few micrometres thick) of biological materials for microscopic examination.

microtubule Minute cylindrical unbranched tubule composed of globular **protein** subunits found either singly or in groups in the **cytoplasm** of **eukaryotic** cells, in which it has the skeletal function of maintaining their shape. Microtubules are also associated with **spindle** formation, and hence are responsible for chromosomal movement during nuclear division.

■ **mid-brain** Part of **brain** that connects the fore-brain to the hind-brain, concerned with processing visual information passed from the fore-brain. Fish, amphibians and birds have a well-developed mid-brain roof, the tectum, which forms the integration centre of their brain. Mammals have a less well-developed mid-brain. Alternative name: mesencephalon. [3.5.e]

■ **middle ear** Air-filled part of the **ear** that is on the inner side of the ear drum and transmits sound waves from the outer ear to the inner ear. Alternative name: tympanic cavity. [3.4.a] [3.5.e] [3.8.a] [14.6.b]

■ **middle lamella** Thin intercellular cementing pectic substances (*see* **pectin**) that hold together adjacent plant cell walls.

■ **mildew** Any fungus disease of plants in which the **mycelium** is visible as pale patches on external surfaces. [2.3.a] [2.3.b] [2.4.b]

milk sugar Alternative name for **lactose**.

■ **milk teeth** First of two sets of teeth possessed by most mammals. Human beings have 20 milk teeth. Alternative name: deciduous teeth. [3.4.b]

■ **mimicry** Close resemblance between one animal and another. The mimicking (in colour, sound, habit or structure) of another of a different species gains advantage by its resemblance to the model; *e.g.* palatable insects that mimic the warning coloration of poisonous species are avoided by their predators. Imitation of, *e.g.*, the background is camouflage, not mimicry. [2.5.d]

■ **mineral salts** Dissolved salts that occur in soil, derived from weathered rock and decomposed plants. They contain essential nutrients for plant growth, which are in turn utilized by herbivores (and the carnivores that in turn feed on them). [3.7.a] [3.8.a] [3.8.b]

■ **miscarriage** Alternative name for a spontaneous **abortion**. [3.4.c]

■ **mitochondrion** Cell organelle in the **cytoplasm** of **eukaryotic** cells, concerned with **aerobic respiration** and hence energy production from the reduction of **ATP** to ADP. Its shape varies from spherical to cylindrical. Large concentrations of mitochondria are observed in areas of high energy consumption, such as muscle tissue. Alternative name: chondriosome. [3.6.a]

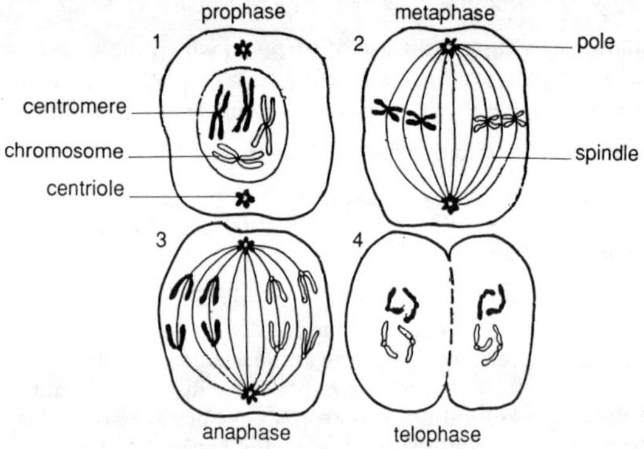

Stages of mitosis

■ **mitosis** The usual type of cell division in which the parent nucleus splits into two identical daughter nuclei, which contain the same number of **chromosomes** and identical **genes** to that of the parent nucleus. Alternative name: karyokinesis. *See also* **meiosis**. [4.7.a]

■ **molar tooth** One of the rearmost teeth of a mammal, used for crushing and grinding food. They are absent from the **milk teeth**, and in carnivores are replaced by **carnassial teeth**. [3.4.b] [3.7.b]

molecular biology Study of biological **macromolecules** (*e.g.* nucleic acids, proteins).

■ **mollusc** Member of the phylum **Mollusca**. [2.3.b] [2.4.a] [2.5.b]

■ **Mollusca** Phylum of invertebrates, all of which have a well-defined body cavity. Molluscs include gastropods, cephalopods, and bivalves (lamellibranchs), such as snails, jellyfish and scallops. Most of the 80,000 species are aquatic, and manufacture a protective shell from dissolved calcium carbonate. [2.3.b] [2.4.a] [2.5.b]

Monera In some classification schemes, the most primitive of the kingdoms of life, consisting of the **prokaryotes** – **cyanophytes** (blue-green algae) and **bacteria**.

■ **mongolism** Alternative name for **Down's syndrome**. [4.6.b] [4.8.b]

monoclonal Derived from a single parent **clone**. [4.9.c]

monoclonal antibody Antibody produced by a single-cell clone, and hence consisting of a single **amino acid** sequence. Such cell clones are produced by the artificial fusion of cancerous and antibody-forming cells from the mouse spleen. The hybrid cells are grown in vitro as clones of cells, with each producing only a single type of antibody molecule.

■ **monocotyledon** Member of one of the two subdivisions of the flowering plants, in which the embryo has a single **cotyledon** (seed-leaf); class Liliopsida. *See also* **dicotyledon**. [2.3.b] [2.4.a] [2.5.b]

■ **monoculture** Describing a form of **agriculture** in which only a single crop is grown continuously.

■ **monocyte** Largest phagocytic **leucocyte**, of order of about

10–12 micrometres. Monocytes have monogranulated cytoplasm with a large oval nucleus.

■ **monohybrid** Offspring of parents that have different **alleles** for a particular gene; one parent has two **recessive** alleles and the other has two **dominant** ones. All offspring inherit one recessive and one dominant allele for the gene. [4.8.a] [4.10.a]

■ **monohybrid cross** Cross between **monohybrids**, giving offspring in the characteristic 3:1 ratio of those with two dominant **alleles** to those with two recessive alleles for a particular gene. [4.8.a] [4.10.a]

■ **monoecious** Describing plants in which separate female and male reproductive bodies are borne on the same individual or flower; *e.g.* maize.

monophyodont Describing an animal that has only one set of (irreplaceable) teeth during its whole life cycle.

monopodium Main axis of a plant stem that undergoes indefinite growth.

■ **monosaccharide** $C_nH_2O_n$ Member of the simplest group of **carbohydrates,** which cannot be hydrolysed to any other smaller units; e.g. the sugars glucose, fructose.

■ **monosodium glutamate** (MSG) White crystalline solid, a sodium salt of the **amino acid** glutamic acid, made from soya bean protein and used as a flavour enhancer. Eating it can cause an allergic reaction in certain susceptible people.

Monotremata Order of egg-laying mammals in the subclass Prototheria found in Australia and New Guinea. There are only three species of monotremes: the duck-billed platypus and two spiny ant-eaters, or echidnas.

monotreme Member of the order **Monotremata**.

■ **morphine** Sedative narcotic **alkaloid** drug isolated from opium, used for pain relief. Alternative name: morphia. [3.4.b] [3.8.b]

morphogenesis Origin and development of form and structure of an organism or part of one.

■ **morphology** Study of the origin, development and structures of organisms. [2.4.a] [2.5.b]

■ **mosaic** *1*. General term for a plant disease caused by viruses that results in patchy leaf coloration. *2*. Organism derived from a single embryo that displays the characteristics of different **genes** in different parts of its body. *See* **chimaera**. *3*. Ordered arrangement that maximizes a functional requirement (*e.g.* positioning of plant leaves that gives maximum exposure to sunlight with a minimum of mutual shading). [2.6.a] [4.9.d]

■ **moss** Plant that belongs to the class **Musci**. [2.3.b] [2.4.a] [2.5.b]

■ **motile** Describing an organism or structure that can move. [3.3.a] [3.4.c]

■ **motor neurone** Nerve cell that transmits impulses from the spinal cord or the brain to a muscle. Alternative names: motor neuron, motor nerve. [3.7.c]

■ **mould** Fungal growth usually consisting of a mass of **hyphae**, especially on rotting food.

■ **moult(ing)** *1*. Periodic shedding of hair or feathers by animals. *2*. Widely used term for ecdysis, the process by which an immature insect sheds its **exoskeleton** in order that it may develop and grow in size.

■ **movement** Change of position of part of the body, a characteristic of living organisms. *See also* **locomotion**.

■ **mucilage** Group of gum-like compounds that produce a slimy solution. Most are highly branched flexible molecules, which may also form part of the cell-wall matrix in plants.

■ **mucin** Any of a number of **glycoproteins** that occur in **mucus**. [2.5.b]

■ **mucous membrane** Moist, **mucus**-lined **epithelium** which itself lines vertebrate internal cavities, including the alimentary, respiratory and reproductive tracts, which are continuous with the outer environment. [3.4.a]

■ **mucus** Slimy substance secreted by the **goblet cells** of **mucous membrane**. It lubricates and protects the epithelial layer on which it is secreted.

■ **mule** Sterile **hybrid** animal born of a mare and a male donkey.

■ **multicellular** Describing plants and animals that have bodies consisting of many cells.

■ **multifactorial inheritance** Existence of more than two **alleles** for one **gene**; *e.g.* as in A, B, O blood grouping. [4.6.b] [4.8.b]

■ **Musci** Class of **spore**-producing plants in the division **Bryophyta**, comprising the mosses. Alternative name: Bryopsida.

■ **muscle** Animal tissue that contracts (by means of muscle fibres) to produce movement, tension and mechanical energy. *See* **involuntary muscle**; **voluntary muscle**. [3.3.a]

mutagen Chemical or physical agent that induces or increases the rate of **mutation**; *e.g.* ethyl methanesulphonate, ultraviolet light, X-rays and gamma-rays. [4.8.c]

■ **mutant** Organism that arises by **mutation**. [4.8.c]

mutarotation Change in the **optical activity** of a solution containing photo-active substances, such as sugars.

■ **mutation** Alteration in the sequence of **bases** encoded by **DNA**, resulting in a permanent inheritable change in the **gene** and consequently the **protein** encoded. Mutations may occur in different ways, and may be induced by a **mutagen** or occur spontaneously. A mutation can be detrimental, *e.g.* those thought to be involved in carcinogenesis (formation of cancer). However, some mutations can be advantageous, *e.g.* in **evolution**, where favourable characteristics may be passed on to offspring. [4.8.c]

■ **mutualism** Relationship between two organisms from which each benefits (*e.g.* cellulose-digesting micro-organisms and the animals, such as ruminants, whose gut they inhabit). *See also* **commensalism; symbiosis**.

■ **mycelium** Mass of **hyphae** that form the body of a **fungus**. [3.6.a]

■ **mycology** Scientific study of **fungi**. [2.5.b]

Mycota Division (phylum) that includes the **fungi**, usually included in the plant kingdom but sometimes accorded a kingdom of its own.

mycotoxin Toxin produced by a **fungus**.

■ **myelin sheath** Thin fatty layer of membranes, produced by **Schwann cells**, that covers the **axon** of most vertebrate **neurones** (nerve cells). [3.7.a]

■ **myeloid tissue Tissue** usually present in **bone marrow** which produces **red blood cells** and other **blood** constituents.

■ **myocardial** Relating to **myocardium**.

■ **myocardium** Muscle tissue of the vertebrate heart.

■ **myoglobin** In vertebrate **muscle** fibre, a **haem** protein capable of binding with one atom of oxygen per molecule. It is abundant in the muscles of diving mammals (*e.g.* seals, whales), in which it acts as an oxygen store.

myology Study of muscles.

■ **myopia** Short-sightedness, a visual defect in which the eyeball is too long (front to back) so that rays of light entering the **eye** from distant objects are brought to a focus in front of the retina. It can be corrected with spectacles or contact lenses made from diverging (concave) lenses. *See also* **hypermetropia**. [15.6.c]

■ **myosin** Fibrous **protein** which, with **actin**, makes up muscle. Movement of myosin fibres between actin fibres causes muscle contraction.

■ **myriapod** Member of the animal class **Myriapoda**. As a general term, myriapod is also taken to include centipedes (class Chilopoda). [2.5.b]

■ **Myriapoda** Class of terrestrial arthropods characterized by a distinct head bearing antennae, mandibles and maxillae, and segmented bodies with many pairs of walking legs; the millipedes.

■ **myxoedema** Disorder caused by lack of hormones from the **thyroid gland**; if present at birth it causes cretinism.

■ **myxomatosis** Disease of rabbits caused by a virus, which has been deliberately introduced in some regions as a form of pest control. [2.5.d] [2.10.a] [3.8.b]

N

■ **nacre** Mother of pearl, the inner layer of the shell of a **mollusc**. It is an iridescent substance, composed mainly of **calcium carbonate**.

NAD (nicotinamide adenine dinucleotide) **Coenzyme** form of the vitamin **nicotinic acid**, necessary in certain enzyme-catalysed oxidation-reduction reactions in cells. Its reduced form is a precursor in the fixation of carbon dioxide in chloroplasts during **photosynthesis**.

■ **nail** Layer of **keratin** that grows on the upper surface of the fingers of human beings and other primates (except tree-shrews, which have claws).

■ **narcotic Analgesic** drug that, in addition to killing pain, causes loss of sensation or loss of consciousness (*e.g.* morphine and other opiates). [3.4.b]

■ **nasal** To do with the nose or sense of smell.

■ **nasal cavity** Cavity located in the head of tetrapods, containing the olfactory sense organs.

■ **nastic movement** Response by plants to stimuli that do not come from any one direction, *e.g.* temperature, humidity. *See also* **tropism**.

■ **natural selection** One of the conclusions drawn by the British naturalist Charles Darwin (1809–82) from the theory of evolution: certain organisms with particular characteristics are more likely to survive and hence pass on their characteristics to their offspring, *i.e.* survival of the fittest. Thus the characteristics of a population are controlled by this process. [4.6.a] [4.9.a]

■ **Neanderthal man** Extinct hominid from the Pleistocene epoch, now usually regarded as a subspecies of *Homo sapiens*. It probably possessed the same size of brain as modern man but with a structure that was different.

■ **nearsightedness** Alternative name for **myopia**. [15.6.c]

■ **nectar** Sticky sweet liquid produced by flowers which attracts insects, small birds and even certain bats (which thus pollinate the flower). It contains up to 80% **sugar** and is used by bees to make honey. [3.4.d]

nekton Aquatic organism that actively swims, as opposed to floating passively; *e.g.* fish, jellyfish and aquatic mammals. *See also* **plankton**.

■ **nematocyst** Small retractable tentacle, possessed by many species of coelenterates, that is used for defence or to catch prey. The hollow tentacle is inflated by a sac of fluid, and may be used to inject venom.

■ **Nematoda** Phylum of invertebrate animals that contains round, thread and eel worms. They have unsegmented bodies that taper at each end. Many nematodes are parasites (*e.g.* filaria). [2.5.b]

■ **nematode** Worm of the phylum **Nematoda**. [2.5.b]

■ **neo-Darwinism** Modern version of the Darwinian theory of **evolution**, expanded to take into account modern knowledge of genetics. *See* **Darwinism**. [4.9.a]

■ **neo-Lamarckism** *See* **Lamarckism**.

neonatal Concerning the newborn.

■ **neoplasm** Tumour or group of cells with uncontrolled growth. It may be benign and localized, or if cells move from their normal position in the body and invade other organs the tumour is malignant. *See also* **cancer**; **metastasis**. [3.8.a] [3.8.b] [4.8.c]

neotenin Insect **hormone** that suppresses the onset of adult characteristics until the final moult.

neoteny Presence of the larval or early stage of development in adulthood. It is important in the evolution of some animal groups.

■ **nephridium** Excretory organ possessed by many invertebrates, consisting of a tube leading from the **coelom** to the external body surface. [3.4.a] [3.5.e] [3.8.a]

■ **nephritis** Inflammation of the **kidney**. [3.8.a]

■ **nephron** Functional filtering unit of the vertebrate **kidney**, consisting of **Bowman's capsule** and the **glomerulus**.

■ **nerve** Structure that carries nervous impulses to and from the central nervous system, consisting of a bundle of **nerve fibres**, and often associated with blood vessels and connective tissue. *See also* **neurone**. [3.7.c]

■ **nerve cell** Alternative name for a **neurone**.

■ **nerve cord** Cord of nervous tissue in invertebrates that forms part of their **central nervous system**.

■ **nerve fibre** Extension of a nerve cell. Nerve fibres may be surrounded by a **myelin sheath** (except at the **nodes of Ranvier**), as in many vertebrates; or they may be unmyelinated and bound by a **plasma membrane**.

■ **nerve gas** Chemical warfare gas that acts on the nervous system, *e.g.* by inhibition or destruction of chemicals (**neurotransmitters**) that transmit a nerve impulse across a **synapse** between nerves.

■ **nerve impulse** Electrical signal conveyed by a **nerve** to carry information throughout the **nervous system**. External stimuli trigger nerve impulses in **receptor** cells, and travel along **afferents** towards the central nervous system. Impulses generated within the **central nervous system** travel along **efferents** towards organs and tissues. Intermediate neurones connect sensory afferent neurones to motor efferent neurones and to the brain.

■ **nervous system** System that provides a rapid means of communication within an organism, enabling it to be aware of its surroundings and to react accordingly. In most animals it consists of a **central nervous system** (CNS) that integrates the sensory input from peripheral nerves which transmit stimuli from receptors (afferents) to the CNS, allowing the appropriate response from the effectors.

■ **neural** Relating to **nerves** or the **nervous system**.

neuroglia Connective tissue between nerve cells (**neurones**) of the brain and spinal cord.

■ **neurone** Basic cell of the **nervous system** which transmits **nerve impulses**. Each cell body typically possesses a nucleus and fine processes: short **dendrites**, an **axon** and a **dendron**. The axon carries impulses to distant effector cells or other neurones, depending on whether it is a sensory or motor neurone. Neurones also make functional contacts over the surface of shorter, thread-like projections from the cell body (dendrites). Alternative names: neuron, nerve cell. [3.7.c]

■ **neurotransmitter** Chemical released by **neurone** endings to either induce or inhibit transmission of nerve impulses across a **synapse**. Neurotransmitters are typically stored in small vesicles near the synapse and released in response to arrival of an impulse. There are more than 100 different types (*e.g.*

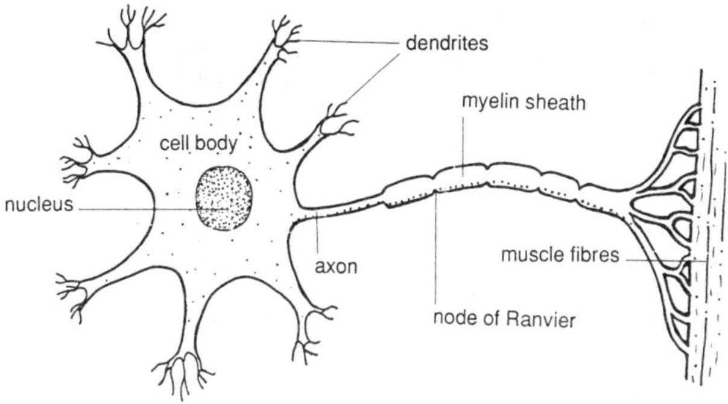

Structure of a neurone

acetylcholine, noradrenalin). Alternative name: transmitter. [3.7.c]

■ **neutralization** Chemical reaction between an **acid** and a **base** in which both are used up; the products of the reaction are a **salt** and water.

neutrophil Type of **leucocyte** (white blood cell), important in the immune system because it ingests bacteria, whose protoplasm can be stained by neutral dyes.

■ **niacin** Vitamin B_3, the only one of the B vitamins that is synthesized by animal tissues. It is used by the body to manufacture the enzyme **NAD**, and a deficiency causes the disease pellagra. Alternative name: nicotinic acid. [3.8.b]

■ **niche** Status or way of life of an organism (or group of organisms) within an environment, which it cannot share indefinitely with another competing organism. Alternative name: ecological niche.

■ **nicotine** Poisonous **alkaloid**, found in tobacco, which potentially binds to the receptor for the **neurotransmitter** acetylcholine. It is used as an insecticide. [3.4.b]

nicotinic acid Alternative name for **niacin**.

■ **nictitating membrane** Protective transparent lid that can be drawn across the eye of many birds and reptiles, and some amphibians and sharks, and which is possessed by a few aquatic mammals, *e.g.* seals.

■ **nitrification** Conversion of **ammonia** and **nitrites** to **nitrates** by the action of nitrifying bacteria. It is one of the important parts of the **nitrogen cycle**, because nitrogen cannot be taken up directly by plants except as nitrates. [3.7.a]

■ **nitrogen cycle** Circulation of nitrogen and its compounds in the environment. The main reservoirs of nitrogen are nitrates in the soil and the gas itself in the atmosphere (formed from nitrates by **denitrification**). Nitrates are also taken up by plants, which are eaten by animals, and after their death the nitrogen-containing **proteins** in plants and animals form ammonia, which **nitrification** converts back into nitrates. Some atmospheric nitrogen undergoes **fixation** by lightning or bacterial action, again leading to the eventual formation of nitrates. [2.7.a]

■ **nitrogen fixation** *See* **fixation of nitrogen**. [2.7.a]

■ **node** *1.* In plants, point of leaf insertion on the shoot axis. *2.* In animals, thickening or junction of an anatomical structure, *e.g.* lymph node (gland), sinoatrial node, node of Ranvier. [3.7.a]

■ **node of Ranvier** One of several regular constrictions along the myelin sheath of a **nerve fibre**. It was named after the French histologist Louis-Antoine Ranvier (1835–1922). [3.7.c]

■ **nodule** Small lump or swelling; *e.g.* nodules containing nitrifying bacteria that form on the roots of certain pulses.

■ **noradrenalin** Hormone secreted by the medulla of the **adrenal glands** for the regulation of the cardiac muscle, glandular tissue and smooth muscles. It is also a **neurotransmitter** in the sympathetic nervous system, where it acts as a powerful vasoconstrictor on the vascular smooth muscles. In the brain, levels of noradrenalin are related to normal mental function, *e.g.* lowered levels lead to mental depression. Alternative name: norepinephrine. [3.9.b]

norepinephrine Alternative name for **noradrenalin**.

normal body temperature Average temperature of the healthy human body, about 37°C.

■ **notochord** Skeletal rod that lies dorsally beneath the **neural tube** in certain stages of development of an **embryo** of a **chordate**. In some chordates it persists in the adult. In **vertebrates** it is replaced by the **spinal column**. [3.4.c]

nucellus Mass of thin-walled cells in the centre of an **ovule** of a plant, containing the megaspore or egg cell. The simple cell of the nucellus becomes the megasporocyte, while the rest acts as a nutritive tissue for the developing megaspore. In some plants it may be retained to form the **endosperm** for the developing embryo, *e.g.* in maize.

■ **nuclear division** *See* **meiosis**; **mitosis**.

nuclear envelope Double membrane that surrounds the **nucleus** of **eukaryotic** cells.

■ **nuclear membrane** Membrane that encloses the **nucleus** of a cell.

nuclease Type of **enzyme** that splits the 'chain' of the **DNA** molecule. Nucleases that act at specific sites are called **restriction enzymes**.

■ **nucleic acid** Complex organic **acid** of high molecular weight consisting of chains of **nucleotides**. Nucleic acids commonly occur, conjugated with **proteins,** as **nucleoproteins,** and are found in cell nuclei and **protoplasm**. They are responsible for storing and transferring the **genetic code**. *See* **DNA** (deoxyribonucleic acid); **RNA** (ribonucleic acid).

nucleoid In a **prokaryotic** cell, the DNA-containing region, similar to the nucleus of a **eukaryotic** cell but not bounded by a membrane.

■ **nucleolus** Spherical body that occurs within nearly all nuclei of **eukaryotic** cells. It is associated with **ribosome** synthesis and is thus abundant in cells that make large quantities of **protein**. It contains protein DNA and much of the nuclear RNA.

nucleoprotein Compound that is a combination of a **nucleic acid** (DNA, RNA) and a **protein**. *E.g.* in **eukaryotic** cells DNA is associated with histones and protamines; RNA in the cytoplasm is associated with protein in the form of the **ribosomes**.

nucleoside Compound formed by partial hydrolysis of a **nucleotide**. It consists of a base, such as purine or pyrimidine, linked to a sugar, such as ribose or deoxyribose; *e.g.* adenosine, cytidine and uridine.

nucleotide Compound that consists of a sugar (ribose or deoxyribose), base (purine, pyrimidine or pyridine) and phosphoric acid. Nucleotides are the basic units from which **nucleic acids** are formed.

■ **nucleus** In biology, the largest cell organelle (about 20 micrometres in diameter), found in nearly all **eukaryotic** cells. It is spherical to oval, containing the genetic material **DNA**, and hence controlling all cell activities. It is surrounded by a double membrane that forms the nuclear envelope. A nucleus is absent from mature mammalian **erythrocytes** (red blood cells) and the mature **sieve-tube** elements of plants.

■ **nutrient** Useful component of food, such as fats, carbohydrates, proteins, vitamins and mineral salts. *See* **diet**.

■ **nutrition** Feeding. The taking in of food is a characteristic of all living organisms, both plants and animals.

nutation In botany, the circular swaying movements made by the growing parts of a plant, *e.g.* a shoot.

nyctinasty Opening and closing of plant organs in response to daily changes of light and temperature. Alternative name: nyctinastic movement.

■ **nymph** Immature stage in the life-cycle of mites, ticks and insects that do not undergo complete **metamorphosis** (*e.g.* a locust). In almost all respects a nymph is a miniature version of the adult, although it usually lacks wings. [2.5.a] [2.6.a]

O

occlusion In biology, closure of an opening (*e.g.* the way an animal's teeth meet when the mouth closes).

■ **ocellus** Simple **eye**, possessed by many invertebrates, consisting of a cluster of receptors that are sensitive to changes in the intensity or colour of light.

Odonata One of the most primitive orders of winged insects, comprising the dragonflies and damselflies.

■ **oesophagus** Muscular tube between the **pharynx** and **stomach** (the gullet), through which food passes by **peristalsis**. [3.4.a] [3.5.e] [3.7.b] [3.8.c]

■ **oestrogen** Female sex **hormone**, a member of a group of **steroid** hormones that act on the sex organs. The most important is oestradiol, which is responsible for the growth and activity of much of the female reproductive system. [3.4.c]

■ **oestrous cycle** Reproductive cycle of most female mammals, during which there is a period of oestrus or 'heat', when ovulation occurs and the female may be successfully impregnated by a male to achieve **fertilization**. The cycle is regulated by hormones. *See also* **menstrual cycle**.

oil *See* **fats and oils**.

oleic acid $C_{17}H_{33}COOH$ **Unsaturated fatty acid** that occurs in many **fats and oils**. It is a colourless liquid that turns yellow on exposure to air, and is used in varnishes.

olfaction Process of smelling. In vertebrates, the incoming nerve impulses from the olfactory sense organs are processed in the olfactory lobes of the brain.

■ **olfactory** Concerning the sense of smell. [3.7.c]

Oligochaeta Class of annelid worms with few chaetae (bristles); *e.g.* earthworms.

oligochaete Worm of the class **Oligochaeta**.

■ **ommatidium** One of the individual elements, composed of a lens linked by nerves directly to the brain, that makes up the **compound eyes** of insects and spiders.

■ **omnivore** Animal that eats both plants and animals. *See also* **carnivore; herbivore**. [2.3.b] [2.4.a] [2.5.b]

oncogenic Cancer-producing. *See also* **carcinogenic**.

oncology Study of **cancer**.

oncovirus Alternative name for **retrovirus**.

ontogeny Developmental history of an organism, from fertilized ovum to sexually mature adult.

oöcyte Cell at a stage of oögenesis before the complete development of an **ovum**. A primary oöcyte undergoes the first meiotic division to give the secondary oöcyte, which after the second meiotic division produces an ovum. *See* **meiosis**.

oögamy Form of sexual reproduction in which the female egg (ovum) is large and non-motile and is fertilized by a small, motile male **gamete**.

oögenesis Origin and development of an **ovum**. *See* **oöcyte**.

■ **oöspore** *1.* Fertilized ovum. *2.* **Zygote** of some algae that has thick walls and food reserves. [3.4.c]

oötecha Egg-case formed by some insects, especially cockroaches and mantids.

operator In biology, region of **DNA** to which a molecule or repressor may bind to regulate the activity of a group of closely linked structural **genes**.

■ **operculum** *1.* Flap that protects the **gill** of bony fishes. *2.* Lid of a moss capsule. *3.* Plate in some **gastropods** that can cover the shell opening.

operon Groups of closely linked structural **genes** which are under control of an operator gene. The operator may be switched off by a repressor, produced by a regulator gene separate from the operon. Another substance, the effector, may inactivate the repressor.

■ **opium** Dried juice from a species of poppy; a bitter nauseous-tasting brown mass with a heavy, characteristic smell. Its narcotic action – for which it is used both medicinally and as an abused drug – depends on the **alkaloids** it contains. *See also* **morphine**. [3.4.b]

■ **optic** Concerning the **eye** and vision. [15.6.b]

optical activity Phenomenon exhibited by some chemical compounds which when placed in the path of a beam of plane-polarized light are capable of rotating the plane of polarization to the left (**laevorotatory**) or right (**dextrorotatory**). Alternative name: optical rotation.

optically active Describing a substance that exhibits **optical activity**.

optical rotation Alternative name for **optical activity**.

■ **optic nerve** Cranial nerve of vertebrates that transmits stimuli from the eye to the brain. [3.4.a] [3.5.e] [3.8.a] [15.6.b]

■ **oral** Concerning the mouth and, in some contexts, speech.

oral hygiene Cleanliness and health of the mouth and teeth.

■ **orbit** In biology, the eye socket.

■ **order** In biological **classification**, one of the groups into which a class is divided, and which is itself divided into families; *e.g.* Lagomorpha (lagomorphs), Rodentia (rodents). [2.3.b] [2.4.a] [2.5.b]

■ **organ** Specialized structural and functional unit made up of various **tissues**, in turn formed of many cells, found in animals and plants; *e.g.* heart, kidney, leaf. [3.4.a] [3.5.e] [3.8.a]

■ **organ culture** Maintenance of an organ in vitro (after removal from an organism) by the artificial creation of the bodily environment.

■ **organelle** Discrete membrane-bound structure that performs a specific function within a **eukaryotic** cell; *e.g.* nucleus, mitochondrion, chloroplast, endoplasmic reticulum. [3.5.a]

■ **organ of Corti** Organ concerned with hearing, located in the **cochlea** of the **ear**. It was named after the Italian anatomist Alfonso Corti (1822–88). [14.6.b]

■ **organ system** Functional unit made up of several **organs**; *e.g.* the digestive system.

Orthoptera Large order of insects, characterized by powerful hindlegs, comprising grasshoppers, crickets, locusts and katydids.

orthotropism Growth straight towards or away from a stimulus; *e.g.* primary roots and shoots are orthotropic to gravity and light, respectively.

■ **osmoregulation** Process that controls the amount of water and electrolyte (salts) concentration in an animal's body. In a saltwater animal, there is a tendency for water to pass out of the body by **osmosis**, which is prevented by osmoregulation by the kidneys. In freshwater animals, osmoregulation by the kidneys (or by **contractile vacuoles** in simple creatures) prevents water from passing into the animal by osmosis. [3.8.a]

■ **osmosis** Movement of a solvent from a dilute to a more concentrated solution across a **semi-permeable** (or differentially permeable) **membrane**. [3.8.a]

■ **osmotic pressure** Pressure required to stop **osmosis** between a solution and pure water. [3.8.a]

■ **ossicle** Alternative name for an **ear ossicle**. [14.6.b]

■ **ossification** Process by which **bone** is formed, especially the transformation of **cartilage** into bone.

Osteichthyes Class of animals that comprises the bony fish.

ostracod Member of the subclass **Ostracoda**.

Ostracoda Subclass of **crustaceans** in which the entire body is enclosed in a smooth, rounded **carapace**; *e.g.* mussel shrimps.

■ **outbreeding** Mating between members of a species that are not closely related. *See also* **inbreeding**.

■ **outer ear** Part of the **ear** that transmits sound waves from external air to the ear drum. [3.4.a] [3.5.e] [3.8.a] [14.6.b]

■ **oval window** Membranous area at which the 'sole' of the stirrup (stapes) bone of the inner **ear** makes contact with the **cochlea**. Alternative name: fenestra ovalis. [3.4.a] [3.5.e] [3.8.a] [14.6.b]

■ **ovarian follicle** Alternative name for the **Graafian follicle**. [3.4.c]

■ **ovary** Female reproductive organ. In vertebrates, there is a pair of ovaries, which produce the ova (eggs) and sex **hormones**. In plants, the ovary is the hollow base of a **carpel** enclosing one or more ovules borne on a placenta; after fertilization the ovary of a plant becomes the **pericarp** of the fruit. [3.4.c]

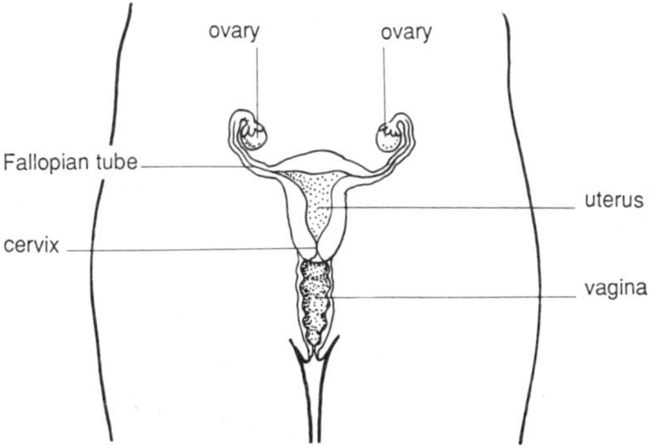

Human ovaries

■ **oviduct** Tube that conducts released ova (eggs) from the **ovaries** after ovulation. Alternative name (in human beings): Fallopian tube. [3.4.c]

■ **oviparity** Animal reproduction in which ova (eggs) are laid by the female either before or after fertilization; *e.g.* as in birds and fish.

■ **ovipositor** Egg-laying structure of female insects, formed from a modified pair of appendages at the hind end of the abdomen (*e.g.* in a parasitic wasp it is a modified sting). Some fish too have ovipositors (*e.g.* bitterling).

■ **ovoviviparity** Animal reproduction in which ova (eggs) hatch inside the female body and the embryo is retained for protection. The embryo obtains its nourishment independently from the yolk store. It occurs in some fish and reptiles.

■ **ovulation** In vertebrates, discharge of an ovum (egg) from a mature **Graafian follicle** at the surface of an **ovary**. In mature human females, ovulation occurs from alternate ovaries at about every 28 days until the **menopause** occurs. [3.4.c]

■ **ovule** Structure in female seed plants. It consists of a nucleus, which contains the embryo sac, surrounded by integuments. After fertilization the ovule develops into the seed. [3.4.d]

ovuliferous scale In cone-bearing plants, woody structure that bears **ovules** and later seeds on its upper surface.

■ **ovum** In animals, unfertilized non-motile female **gamete** produced by the **ovary**. Alternative names: egg cell, egg. [3.4.c]

oxidase Collective name for a group of **enzymes** that promote **oxidation** within plant and animal cells.

■ **oxidation** Combination of a substance with **oxygen**. It may occur rapidly (as in combustion and aerobic respiration) or slowly (as in rusting and other forms of corrosion). Oxidation is one of the causes of spoilage in food, sometimes combated by using **additives** called anti-oxidants.

■ **oxidizing agent** Substance that causes **oxidation**. Alternative name: electron acceptor.

■ **oxygen** Gas that is essential for life in all living organisms (except for a few lower forms such as certain bacteria). In animal tissues its main function is to oxidize **glucose** to release energy. *See* **aerobic respiration**; **respiration**. [3.6.a]

■ **oxygen debt** Physiological condition that induces **anaerobic respiration** in an otherwise **aerobic** organism. It occurs during anoxia, caused *e.g.* by violent exercise. [3.6.a]

■ **oxyhaemoglobin** Product of **respiration** formed by the combination of **oxygen** and **haemoglobin**. [3.5.b] [3.6.a]

■ **ozone** O_3 Allotrope of oxygen that contains three atoms in its molecule. It is formed from oxygen in the upper atmosphere by the action of ultraviolet light, where it also acts as a shield that prevents excess ultraviolet light reaching the Earth's surface. [5.5.a] [5.9.a]

■ **ozone layer** Layer in the upper atmosphere at a height of between 15 and 30 km, where **ozone** is found in its greatest concentration. It filters out ultraviolet radiation from the Sun which would otherwise be harmful. If this layer were destroyed, or depleted to a great extent, life on Earth would be endangered. Alternative name: ozonosphere. [5.9.a]

P

pachytene Stage in **prophase** or first division of **meiosis**, in which the paired **chromosomes** shorten and thicken, appearing as two **chromatids.**

paedogenesis Sexual reproduction in some animals by the immature forms (larvae and pupae).

■ **palaeontology** Study and interpretation of **fossils**.

palisade Main photosynthetic tissue of a leaf.

■ **palp** Jointed sensory structure located on the mouthparts of insects and other invertebrates.

■ **pancreas** Gland situated near the **duodenum** that has digestive and endocrine functions. The enzymes amylase, trypsin and lipase are released from it during digestion. Special groups of cells (the islets of Langerhans) produce the **hormones insulin** and **glucagon** for the control of blood sugar levels. [3.4.a] [3.5.e] [3.7.b] [3.8.a]

■ **pancreatic juice** Liquid containing digestive **enzymes** that is secreted by the **pancreas** and passed to the **duodenum**. [3.4.a] [3.5.e] [3.7.b] [3.8.a]

■ **pandemic** Describing a disease that affects people or animals throughout the world. *See also* **endemic**; **epidemic**. [3.8.a] [3.8.c] [4.9.a]

panicle Branched **racemose inflorescence**, common in grasses.

pantothenic acid Constituent of **coenzyme A**, a carrier of **acyl groups** in biochemical processes. It is required as a B **vitamin** by many organisms, including vertebrates and yeast.

papain Proteolytic **enzyme**, which digests proteins, found in various fruits and used as a meat tenderizer.

papovavirus Member of a group of double-stranded DNA viruses which infect the cells of higher vertebrates, in which they can cause tumours.

■ **parasite** Organism that is the beneficial partner in **parasitism**. *See also* **ectoparasite; endoparasite**. [3.7.c]

■ **parasitism** Intimate relationship between two organisms in which one (the parasite) derives benefit from the other (the host), usually to obtain food or physical support. Parasitism can have minor or major effects on the survival of the host. *See also* **commensalism; symbiosis.**

■ **parasympathetic nervous system** Branch of the **autonomic nervous system** used in involuntary activities, for which **acetylcholine** is the transmitter substance. Effects of the parasympathetic nervous system generally counteract those of the **sympathetic nervous system**.

■ **parathyroid** Four **endocrine glands** embedded in the **thyroid** in the neck which release a **hormone** that controls the levels of calcium in the blood.

■ **parenchyma** Plant tissue composed of round tightly packed cells, used as packing tissue and for the storage of **starch**.

■ **parthenogenesis** Development of a new individual from an unfertilized egg, which occurs in certain groups of invertebrates. Eggs may be formed by **mitosis** instead of **meiosis** and are **haploid**. In haploid parthenogenesis, the organism produced is also haploid, *e.g.* drones in the honeybee. [3.4.c]

■ **parturition** Birth of a full-growth **foetus** at the completion of pregnancy (gestation). [3.4.c]

■ **pasteurization** Process of heating food or other substances under controlled conditions. It was developed by the French chemist Louis Pasteur (1822–95) to destroy **pathogens**. It is widely used in industry, *e.g.* milk production. [7.5.b]

■ **patella** Bone in front of the knee joint. Alternative name: kneecap. [3.3.a]

■ **pathogen** Disease-causing organism, commonly used of micro-organisms such as **bacteria** and **viruses**. [3.8.b] [7.5.b] [7.7.b]

■ **pearl** Lustrous, often spherical, accretion formed from the layering of **nacre** (calcium carbonate) on a foreign particle inside the shells of certain molluscs, *e.g.* oysters.

■ **pectin** Complex **polysaccharide** derivative present in plant cell walls, to which it gives rigidity. It can be converted to a gel form in sugary acid solution and is used commercially to help to make jam set.

■ **pectoral** Concerning the part of the front end of a vertebrate's body which supports the shoulders and forelimbs. [3.3.a]

■ **pedicel** Stem that attaches an individual flower to the main flower stem (peduncle). [3.4.a]

■ **peduncle** Main flower stem of a plant (*see also* **pedicel**).

■ **pelagic** Describing fish that normally occur in free-swimming shoals in waters less than 200m deep. *See also* **benthic**.

■ **pelvic** Concerning the part of the rear end of a vertebrate's body which supports the hindlimbs. [3.3.a]

■ **pelvis** *1.* Part of the skeleton (pelvic girdle) to which a vertebrate's hindlimbs are joined. *2.* Cavity in the **kidney** which receives urine from the tubules and drains it into the **ureter**. [3.3.a] [3.4.a] [3.4.c]

■ **penicillin** Member of a class of **antibiotics** produced by moulds of the genus *Penicillium*. It inhibits growth of some **bacteria** by interfering with cell-wall biosynthesis. [3.8.b]

■ **penis** Copulatory organ possessed by males or hermaphroditic forms in some animals, which transfers sperm to the female to achieve internal fertilization. In mammals it also houses the **urethra**, through which urine is discharged. [3.4.a] [3.4.c] [3.5.a] [3.8.a]

■ **pentadactyl** Having five digits. It describes the limb typical of most tetrapod vertebrates.

pentose Monosaccharide carbohydrate (sugar) that contains five carbon atoms and has the general formula $C_5H_{10}O_5$; *e.g.* **ribose** and **xylose**. Alternative name: pentaglucose.

■ **pepsin Enzyme** produced in the stomach which, under acid conditions, brings about the partial **hydrolysis** of **polypeptides** (thus helping in the digestion of **proteins**).

peptidase Enzyme, often secreted in the body (*e.g.* by the intestine), which degrades **peptides** into free **amino acids**, thus completing the digestion of **proteins**.

peptide Organic compound that contains two or more **amino acid** residues joined covalently through peptide bonds (–NH–CO–) by a condensation reaction between the carboxyl group of one amino acid and the amino group of another. Peptides polymerize to form **proteins**.

■ **perennial** Plant that lives for a number of years and may be woody (*e.g.* trees and shrubs) and continuously grow or have herbaceous stems which die at the end of each season and are replaced. *See also* **annual**; **biennial**. [2.3.c]

perennating organ Part of a plant that enables it to survive over the winter; *e.g.* a rhizome or tuber.

■ **perianth** Outer part of a flower that surrounds the sexual organs, usually consisting of the **sepals** and **petals**. [3.4.d]

■ **pericarp** Fruit coat formed from the **ovary** wall in plants after fertilization. [3.4.d]

■ **perilymph** Fluid that surrounds the **cochlea** of the **ear**. [14.6.b]

■ **peripheral nervous system** That part of the nervous system that does not include the **central nervous system** (CNS). [3.7.c]

perissodactyl Having an odd number of toes; a member of the **Perissodactyla**.

Perissodactyla Order of **ungulate** mammals whose members have an odd number of toes (one or three), including horses, rhinoceroses and tapirs. *See also* **Artiodactyla**.

■ **peristalsis** Involuntary waves of muscular contraction produced along tubular structures in the body, *e.g.* in the oesophagus or intestine to push food along. [3.7.b]

■ **permanent teeth** Second set of teeth used by most mammals in adult life (after they have displaced the first set of milk or deciduous teeth).

■ **permeable** Porous; describing something (*e.g.* a membrane) that exhibits **permeability**.

■ **permeability** *1.* Rate at which a substance diffuses through a porous material. *2.* Extent to which a substance can pass through a membrane. Membranes may be semi-permeable, *e.g.* plasma membrane, which allows small molecules such as those of water to pass through.

■ **perspiration** Watery fluid produced by sweat glands which has a cooling effect as it evaporates from the surface of the skin. Alternative name: sweat.

■ **pesticide** Compound used in agriculture to destroy organisms that can damage crops or stored food, especially insects and rodents. Pesticides include **fungicides**, **herbicides** and **insecticides**. The effects of some of them, *e.g.* organic chlorine compounds such as DDT, can be detrimental to the ecosystem. [2.6.b] [4.9.c]

pet Animal that is kept in the home for company or pleasure.

■ **petal** Part of a flower that makes up the **corolla**, often scented and with a bright colour (to attract pollinating insects). Petals are regarded as modified leaves. [3.4.a] [3.4.d] [3.5.e] [3.8.a]

■ **Petri dish** Sterilizable circular glass plate with a fitted lid used in microbiology for holding media on which micro-organisms may be cultured. It was named after the German bacteriologist Julius Petri (1852–1921).

■ **pH** Hydrogen ion concentration (grams of hydrogen ions per litre) expressed as its negative logarithm; a measure of acidity and alkalinity. For example, a hydrogen ion concentration of 10^{-3} grams per litre corresponds to a pH of 3, and is acidic. A pH of 7 is neutral; a pH of more than 7 is alkaline. [3.8.a]

phage *See* **bacteriophage**.

■ **phagocyte** Cell that exhibits **phagocytosis**. It is employed in the defence against invasion by foreign organisms, *e.g.* macrophages in human beings. [3.5.b] [3.8.b]

■ **phagocytosis** Engulfment of external solid material by a cell, *e.g.* a **phagocyte**. It is also the method by which some unicellular organisms (*e.g.* protozoa) feed.

■ **phalanges** Bones in the digits of the hand or foot. [3.5.b]

■ **pharmacology** Study of the properties, manufacture and reactions of drugs.

■ **pharmacy** Preparation and dispensing of drugs, and the place where this is done.

■ **pharynx** Area that links the buccal cavity (mouth) to the oesophagus (gullet) and the nares (back of the nostrils) to the trachea (windpipe). Food passes via the pharynx to the oesophagus when the **epiglottis** closes the entrance of the trachea. [3.4.a] [3.5.e] [3.7.b] [3.8.a]

phasmid Member of a group of insects that are characterized by extreme elongation of the body and highly effective camouflage, *e.g.* stick insects and leaf insects.

phellem Alternative name for **cork**.

phellogen Layer of **cambium** tissue below the bark of a woody plant. Alternative name: cork cambium.

■ **phenotype** Outward appearance and characteristics of an organism, or the way **genes** express themselves in an organism. Organisms of the same phenotype may possess different **genotypes**; *e.g.* in a **heterozygous** organism two **alleles** of a gene may be present in the genotype (genetic make-up) with the expression of only one in the phenotype (appearance). [4.6.a] [4.8.a]

phenylalanine $C_6H_5CH_2CH(NH_2)COOH$ Essential amino acid that possesses an **benzene ring**.

■ **pheromone** Chemical substance produced by an organism which may influence the behaviour of another. *E.g.* in moths, pheromones act as sexual attractants; in social insects such as bees, pheromones have an important role in the development and behaviour of the colony. [4.6.a]

■ **phloem** Type of vascular tissue present in plants, which consists mainly of living cells or **sieve tubes** used for the transport of food material from the leaves to other areas of the plant. *See also* **xylem**. [3.4.a] [3.5.e] [3.8.a]

phosphatidyl choline Alternative name for **lecithin**, a **phospholipid** constituent of **plasma membranes**.

■ **phospholipid** Member of a class of complex **lipids** that are major components of cell **membranes**. They consist of molecules containing a phosphoric(V) acid **ester** of **glycerol** (*i.e.* phosphoglycerides), the remaining **hydroxyl groups** of the glycerol being esterified by **fatty acids**. Alternative names: phosphoglyceride, phosphatide, glycerol phosphatide.

■ **phosphorylation** Process by which a **phosphate** group is transferred to a molecule of an organic compound. In some substances, *e.g.* **ATP**, a high-energy bond may be formed by phosphorylation, which is essential for energy-transfer in living organisms. It is an important biochemical end-reaction which modifies the conformation of molecules such as enzymes, receptors, etc.

■ **photochemical reaction** Chemical reaction that is initiated by the absorption of light. The most important phenomenon of this type is **photosynthesis**. It is also the basis of photography. [3.6.b] [3.7.a] [3.9.a]

photolysis Photochemical reaction that results in the decomposition of a substance.

■ **photoperiodism** Influence of day and night on the activities of an organism. *E.g.* the flowering of **plants** is controlled by the photoperiod and regulated by **phytochrome**. [3.6.b]

photophosphorylation Process during **photosynthesis** that results in the formation of **ATP** from energy derived from sunlight via **chlorophyll**. The reactions also produce **hydrogen**, which is used in combination with **carbon dioxide** to make **sugars**. Alternative names: photosynthetic phosphorylation. *See also* **phosphorylation**; **photosynthesis**. [3.6.b] [3.7.a] [3.9.a]

■ **photoreceptor Receptor** consisting of sensory cells that are stimulated by light, *e.g.* light-sensitive cells in the **eye**.

■ **photosynthesis** Type of **autotrophic** nutrition employed by green plants which involves the synthesis of organic compounds (mainly **sugars**) from **carbon dioxide** and water. Sunlight is used as a source of energy, which is trapped by **chlorophyll** present in **chloroplasts**. The process consists of a light stage, in which energy is converted into **ATP** and water is split into **hydrogen** and **oxygen**. Hydrogen is subsequently combined with carbon dioxide in the dark stage to form **carbohydrates**. Photosynthetic **bacteria** use different sources of hydrogen in the process. [3.6.b] [3.7.a] [3.9.a]

■ **photosynthetic pigment Pigment** involved in the trapping of light energy during **photosynthesis**. The chief photosynthetic pigments are **chlorophylls**. Other accessory pigments also trap energy, *e.g.* **carotenoids**. [3.6.b] [3.7.a] [3.9.a]

■ **phototaxis** Movement of an organism in response to stimulation by light.

■ **phototropism** Growth movement or **tropism** that occurs in response to light. Plant stems are positively phototropic and grow towards light, whereas roots are negatively phototropic. The response results from the action of **auxins**. [3.6.b] [3.7.a] [3.9.a]

phthalocyanine Member of an important class of synthetic organic dyes and pigments. They are blue to green and used for colouring paints, printing inks, synthetic plastics and fibres, rubber, etc.

phylogeny Relationship between groups of organisms (*e.g.* members of a **phylum**) based on the closeness of their evolutionary descent.

- **phylum** In biological **classification**, one of the groups into which the animal kingdom is divided. The members of the group, although often quite different in form and structure, share certain common features; *e.g.* the phylum Arthropoda (arthropods) includes all animals with jointed legs and an exoskeleton. Phyla are subdivided into classes. The equivalent of a phylum in the plant kingdom is a division. [2.3.b] [2.4.a] [2.5.b]

- **physiology** Branch of biology that is concerned with the functioning of living organisms (as opposed to anatomy, which deals with their structure).

phytoalexin Substance produced by plants that prevents the growth of some micro-organisms (*e.g.* fungi) on them.

phytochrome Light-sensitive pigment present in small quantities in plants. When activated by light of a specific wavelength (660nm), it functions as an **enzyme** to initiate growth reactions, including the development of stems, roots and leaves, germination, flowering and the formation of other pigments. The activated form of phytochrome is either gradually lost or it may be reconverted to its original inactive form in darkness or in red light (of wavelength 730nm).

- **phytoplankton** Microscopic **algae** that float or drift on the surface waters of ponds, lakes and seas; part of the **plankton**. It forms a major source of food for fish and whales, and for this reason is of great ecological and economic importance. *See also* **zooplankton**.

- **pigment** Name given to some naturally occurring coloured substances; *e.g.* green **chlorophyll** in plants and red **haemoglobin** in blood.

- **pileus** Cap of a fungus which has the form of a mushroom or toadstool.

■ **piliferous layer** Part of the **epidermis** of a plant's root that bears or produces hair-like structures (root hair cells). [3.7.a]

■ **pineal gland** Club-shaped, elongated outgrowth from the roof of the vertebrate forebrain. It may act as a third eye in some lower bony fishes; in other vertebrates it serves as a **hormone**-producing organ whose secretory function is regulated by light entering the body via the eyes. In human beings its role is not clear. Alternative names: pineal body, epiphysis. [3.9.b]

■ **pinna** *1.* In mammals, the part of the **ear** that extends beyond the skull, consisting of a cartilaginous flap. It covers and protects the opening of the ear, and in some animals (*e.g.* dogs, horses, elephants) it may be moved independently of the head to help the animal to ascertain the direction of sounds. Alternative name: auricle. *2.* In birds, a wing or feather. *3.* In fishes, a fin. [3.4.a] [3.5.e] [3.8.a] [14.6.b]

pinocytosis Uptake of particles and macromolecules by living cells. Alternative name: endocytosis.

■ **Pisces** Superclass of animals that comprise the fish, divided into the classes **Agnatha, Elasmobranchii** (Chondrichthyes) and **Osteichthyes**. [2.3.b] [2.4.a] [2.5.b]

■ **pituitary Endocrine gland** situated at the base of the brain in vertebrates, responsible for the production of many **hormones** and thus the major controller of the endocrine system. The anterior (front) lobe produces **growth hormone, luteinizing hormone, follicle-stimulating hormone, thyroid-stimulating hormone**, lactogenic hormone and **ACTH**. The posterior (rear) lobe secretes **oxytocin** and **vasopressin** produced in the **hypothalamus**. Alternative names: pituitary gland, hypophysis. [3.7.c] [3.9.b]

■ **placenta** *1.* In mammals, vascular organ that attaches the

foetus to the wall of **uterus** (womb). *2.* In plants, the vascular part of the **ovary** to which the **ovules** are attached. [3.4.c]

■ **plankton** Microscopic organisms that live at the surface of seas and lakes. They consist of animals (**zooplankton**) and plants (**phytoplankton**), and are important as food for animals as diverse as insects and whales. [2.6.a]

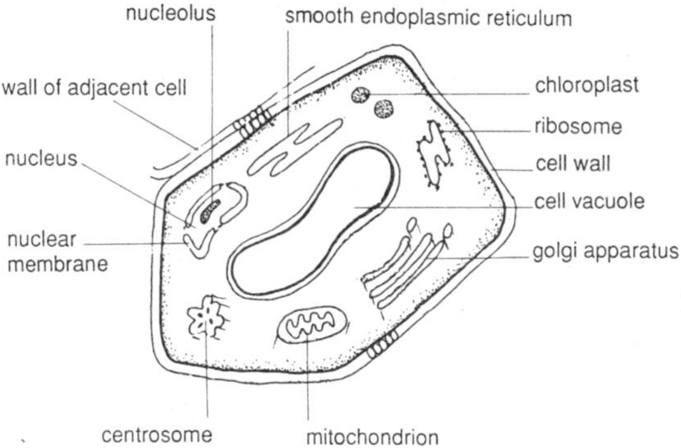

A plant cell

■ **plant** Member of a kingdom (Plantae, or Metaphyta) of mostly **autotrophic** organisms that possess a cell structure differing from **animal** cells. Plant cells contain a **cellulose** wall, which gives them a rigid structure. Green plants contain the pigment **chlorophyll** for carrying out **photosynthesis**, by means of which organic compounds may be generated from inorganic constituents. Members of the plant kingdom vary greatly, from single-celled to multicellular organisms, and they inhabit a wide range of **habitats**. [2.3.b] [2.4.a] [2.5.b]

■ **plaque** Film that accumulates on teeth, consisting of food debris and bacteria, which becomes hardened to form a calculus. The bacteria produce acid as a waste product of their metabolism, and the acid attacks the enamel of the teeth, causing **dental caries** (decay). Alternative name: tartar. [3.4.b]

■ **plasma** Colourless fluid portion of **blood** or **lymph** from which all cells have been removed. [3.5.b] [3.5.c] [3.8.a.]

plasma cell Large egg-shaped cell with granular, basophilic **cytoplasm** except for a clear area around the small eccentrically placed nucleus. Its function is believed to be **antibody** synthesis. Alternative name: plasmacyte.

■ **plasma membrane** Thin layer of tissue consisting of **fat** and **protein** that forms a boundary surrounding the **cytoplasm** of **eukaryotic** cells and its organelles. It is a differentially **permeable** membrane that separates adjacent cells and cavities. Alternative name: cytoplasmic membrane. [3.8.a]

plasma protein **Protein** in the **plasma** of blood (*e.g.* **antibodies** and various **hormones**).

plasmasol Alternative name for **endoplasm**.

■ **plasmolysis** In plants, shrinking of **cytoplasm** away from the cell wall resulting from the loss of water from the central vacuole when the cell is placed in an osmotically more concentrated solution than its cell sap. [3.4.a] [3.5.e] [3.7.a] [3.8.a]

■ **plastid** One of a group of DNA-containing organelles found in plant cells, *e.g.* **chloroplasts**. Plastids perform a number of different functions, and are commonly classified into chromoplasts, which contain pigments, and leucoplasts, which do not.

■ **platelet** Small non-nucleated oval or round fragment of cells from the red bone marrow found in mammalian **blood**. There are about 250,000 to 400,000 per mm^3 in human blood, which are required to initiate blood clotting by disintegrating and liberating thrombokinase. In some vertebrates platelets are represented by thrombocytes, which are small spindle-shaped nucleated cells. [3.5.b] [3.8.a]

Platyhelminthes Phylum of simple invertebrates; flatworms. Parasitic platyhelminths include flukes and tapeworms; *Planaria*, a turbellarian, is a non-parasitic type.

■ **pleura** Double membrane that covers the lungs and lines the chest cavity, with fluid between the membranes. Alternative name: pleural membranes. [3.6.a]

■ **pleural** To do with the lungs.

■ **plexus** Network of interlacing nerves.

■ **plumule** *1.* In birds, small soft feather that forms down. *2.* In seed plants, small leafy part of an embryonic shoot. [3.7.a]

Pogonophora Phylum of simple worm-like marine animals; beardworms.

■ **poikilothermic** Describing an animal that is 'cold-blooded' and relies on the heat of the environment to warm its body; *e.g.* all invertebrates and, among vertebrates, fish, amphibians and reptiles. *See also* **homoiothermic**. [2.4.a]

polar body One of a pair of minute cells that divide off from an **ovum** (egg) when the **oöcyte** undergoes **meiosis**. Alternative names: polocyte, polar globule, directive body.

■ **pollen** Dust-like microspore of a seed plant produced by microsporangium cones in **gymnosperms** and by anthers in **angiosperms**. Each grain contains male **gametes**. If these gametes are carried by an external agent such as wind, insects or water to the **ovules** of gymnosperms or to the **stigma** of angiosperms, they produce fertilization. In susceptible (*i.e.* sensitive) people, pollen can be a powerful **antigen** that results in a vigorous allergic response (*see* **allergy**). Alternative name: farina. [3.4.a] [3.5.e] [3.7.a] [3.8.a]

■ **pollen sac** In **angiosperms**, four sacs (locules) of an **anther** in which **pollen** is produced. In conifer species the microspore of the male **strobilus** bears a variable number of pollen sacs. [3.4.a] [3.5.e] [3.7.a] [3.8.a]

■ **pollen tube** Fine filamentous process developed from **pollen** grains after they become attached to a **stigma**. It grows towards the **ovule**, directing the male nuclei to the ovule in the embryo sac, where fertilization occurs. [3.4.d]

■ **pollination** Transference of **pollen** prior to fertilization by agents such as wind, insects, birds, water etc. from an **anther** to a **stigma** in **angiosperms** and from the male to the female cone in **gymnosperms**. [3.4.a] [3.5.e] [3.8.a]

■ **pollution** Harmful changes or presence of undesirable substances in the environment which result from mankind's industrial or social activities. Pollution of the atmosphere includes the presence of sulphur dioxide, which causes acid rain, and of chlorofluorocarbons (CFCs), which have been linked to the depletion of the **ozone layer** in the stratosphere. River pollution is caused mainly by agricultural run-off (*e.g.* of fertilizers or slurry) or discharge of chemicals. Pollution of the oceans is caused by oil spillage or dumping of untreated sewage, industrial wastes or chemical wastes. [5.5.a]

Polychaeta Class of **annelid** worms that have many chaetae (bristles). Common polychaetes are the bristleworms and ragworms.

polychaete Worm of the class **Polychaeta**.

■ **polymorphism** Occurrence of an organism in more than one structural form during its life cycle. [4.6.a]

polynucleotide Polymer of many **nucleotides**.

■ **polyp** *1*. Body type of **Anthozoa** and **Hydrozoa** (*e.g.* corals and hydra), which may reproduce asexually by budding or splitting and sexually to give rise to new polyps or **medusae**. *2*. In some other **Coelenterata**, a sedentary stage in their life cycle in which the body is cylindrical, with a mouth surrounded by tentacles at one end and attached to a fixed surface at the other end. *3*. **Polypus**.

■ **polypeptide** Chain of **amino acids** which is a basic constituent of **proteins**. It may be broken down by **enzyme** action (digestion) to form **peptides**. The functionally significant linking and folding of polypeptides makes up the three-dimensional structure of a protein. [3.7.b]

■ **polyploidy** Condition in which a cell or organism has three to four times the normal **haploid** or gametic number. It is often made use of in plant breeding because it results in the production of larger and more vigorous crops. Because it disturbs the sex-determining mechanism, polyploidy is rare in animals, and would result in sterility. [4.6.a]

■ **polypus** Pendulous but usually benign **tumour** that grows from mucous membrane (*e.g.* in the nose or womb). Alternative name: polyp.

■ **polysaccharide** High molecular weight **carbohydrate**, linked by **glycoside** bonds, that yields a large number of **monosaccharide** molecules (*e.g.* simple sugars) on **hydrolysis** or **enzyme** action. The most common polysaccharides have the general formula $(C_6H_{10}O_5)_n$; e.g. starch, cellulose, etc. [3.7.b]

■ **pome** False fruit with a fleshy part which develops from the flower's receptacle and not (as in a true fruit) its ovary; *e.g.* apple, pear. [3.4.d]

pons Thick bundle of nerve fibres that relays impulses between different parts of the brain. It joins the medulla oblongata to the midbrain.

population *1.* Human inhabitants of a country. *2.* Animals or plants in a given area.

population genetics Study of the theoretical and experimental consequence of Mendelian inheritance on population levels, taking into account the **genotypes, phenotypes, gene** frequencies and the mating systems. [4.9.d]

■ **pore** Any small opening; *e.g.* pores in the skin through which perspiration passes from sweat glands.

■ **porosity** Property of substance that allows gases or liquids to pass through it.

porphyrin Member of an important class of naturally occurring organic **pigments** derived from four **pyrrole** rings. Many form **complexes** with metal ions, as in *e.g.* chlorophyll, haeme, cytochrome, etc.

■ **portal vein** Any vein connecting two capillary networks, thus allowing for blood regulation from one network by the other; *e.g.* the hepatic portal vein connects the intestine with the liver. [3.4.a] [3.5.e] [3.7.b] [3.8.a]

■ **posterior** In bilaterally symmetrical animals, the end of the body directed backwards during locomotion; the rear or hind end. In bipedal animals (*e.g.* human beings), it corresponds to the **dorsal** side of quadrupeds.

poxvirus Member of a group of large DNA-containing viruses that are responsible for smallpox, cowpox and some animal tumours.

■ **precursor** Intermediate substance from which another is formed in a chemical reaction. *E.g.* blood clotting is dependent on precursors present in the blood but activated only when blood is exposed to air in wounding. [3.5.b] [3.8.b]

■ **predator** Animal that hunts and eats other animals (the prey). [2.5.d] [2.10.a] [2.10.b]

■ **pregnancy** Time that elapses between fertilization or implantation of a fertilized ovum and an animal's birth; the time that an animal spends as an embryo or foetus. Alternative name: gestation. The time that an embryo reptile or bird spends in an egg between laying and hatching is usually termed **incubation**. [3.4.c]

■ **premolar** Grinding and chewing tooth located behind the **canine teeth** and in front of the **molar teeth**. An adult human being has eight premolars, two in each side of each jaw. [3.4.b] [3.5.e] [3.7.b]

■ **presbyopia** Age-related loss of **accommodation** of the human **eye**. Loss of elasticity of the eye lens makes it difficult to focus on near objects. *See also* **hypermetropia**. [15.6.c]

■ **prey** Animal that is hunted for food by a **predator**. [2.5.d] [2.10.a] [2.10.b]

■ **primary growth** Growth of roots and shoots derived from the apical **meristem**, which gives rise to the primary plant body. Alternative name: apical growth.

■ **primary sexual characteristics** Sexual features that are present from birth – *i.e.* excluding **secondary sexual characteristics**. [3.6.c]

■ **Primates** Order of mammals that includes tarsiers, pottos, lemurs, monkeys, great apes and man. They are characterized in evolutionary terms by maintaining a generalized limb structure, increasing digital mobility, binocular vision and progressive development of the cerebral cortex of the brain. Primate young undergo a long period of growth and development, during which they learn from their parents. [2.3.b] [2.4.a] [2.5.b]

primordium Collection of cells that gives rise to a tissue, organ or a group of associated organs; *e.g.* apical shoot and apical root primordia in plants (*i.e.* at the apex of the growing shoot and root).

■ **proboscis** Tube-like organ of varying form and functions. In insects, a proboscis is a filamentous structure that projects outwards from the mouthparts, functioning as a piercing and sucking device for obtaining liquid food. In elephants (order Proboscidea), the proboscis is the trunk, and in some marine animals it is a tube-like pharynx that can be protruded.

■ **producer** Plant in a **food web** that uses **photosynthesis** to convert light energy to chemical energy (which can be used as food by animals).

proenzyme Alternative name for **zymogen**.

profile In an ecological survey, method of recording details of the land in terms of the contours and height or depth of vegetation.

■ **progesterone Steroid sex hormone** secreted by the **corpus luteum** of the mammalian ovary, placenta, testes and adrenal cortex. In females it prepares the **uterus** for the implantation of a fertilized ovum (egg) and during **pregnancy** maintains nourishment for the embryo by developing the placenta, inhibiting ovulation and menstruation, and stimulating the growth of the mammary glands.[3.4.c]

prokaryote DNA-containing, single-celled organism with no proper **nucleus** or **endoplasmic reticulum**; *e.g.* bacteria, blue-green algae (cyanophytes). *See also* **eukaryote**.

prokaryotic Describing or resembling a **prokaryote**.

prolactin Protein **hormone** secreted by the anterior **pituitary**. In mammals it stimulates lactation and promotes functional activity of the **corpus luteum**. Alternative names: luteotrophin, mammary stimulating hormone, mammogen hormone, mammotrophin.

proline White crystalline **amino acid** that occurs in most **proteins**.

promoter Substance used to enhance the efficiency of a **catalyst**. Alternative name: activator. *See also* **coenzyme**.

■ **propagation** *1*. In botany, any form of plant reproduction, especially when manipulated by human beings in gardening and agriculture. [3.4.d]

■ **prophase** First stage of **cell division** in **meiosis** and **mitosis**. During prophase **chromosomes** can be seen to thicken and shorten and to be composed of **chromatids**. The **spindle** is assembled for division of chromosomes and the **nuclear membrane** disintegrates. In meiosis the first prophase is extended into several stages. [4.6.a] [4.7.a]

prostaglandin Member of a group of **unsaturated** fatty acids that contain 20 carbon atoms. They are found in all human tissue, and particularly high concentrations occur in semen.

Their activities affect the nervous system, circulation, female reproductive organs and metabolism. Most prostaglandins are secreted locally and are rapidly metabolized by **enzymes** in the tissue.

■ **prostate gland** Gland located at the base of the urinary bladder that forms part of the reproductive system of male mammals. The size of the gland and the quantity of its secretion are controlled by **androgens**. Its function is secretion of a fluid containing **enzymes** and antiglutinating factor, which contributes to the production of semen. [3.4.a] [3.5.e] [3.8.a]

prosthetic group Non-protein portion of a conjugated **protein**, *e.g.* haem group in haemoglobin.

■ **protease Enzyme** that breaks down **protein** into its constituent **peptides** and **amino acids** by breaking peptide linkages (*e.g.* pepsin, trypsin).

■ **protein** Member of a class of high molecular weight **polymers** composed of a variety of **amino acids** joined by **peptide** linkages. In conjugated proteins, the amino acids are joined to other groups. Proteins are extremely important in the physiological structure and functioning of all living organisms because the greater part of protoplasm and all enzymes are proteins.

■ **protein synthesis** Process by which **proteins** are made in cells. A molecule of **messenger RNA** decodes the sequence of copied DNA on **ribosomes** in the cytoplasm. A **polypeptide** chain is generated by the linking of **amino acids** in an order instructed by the base sequence of **messenger RNA**.

■ **prothallus** In pteridophytes (ferns), the **gametophyte** generation which consists of a flattened, free-living **haploid** disc of cells bearing sex organs. Homosporous plants produce only one type of prothallus, which bears both the male and female sex organs. Heterosporous plants produce two different types of prothalli, a male prothallus which develops **antheridia** and a female one which develops **archegonia**.

protist Member of the **Protista**. [2.5.b]

■ **Protista** Kingdom that contains simple organisms such as **algae**, **bacteria**, **fungi** and **Protozoa**, although sometimes multicellular organisms are excluded. [2.5.b]

■ **protoplasm** Usually transparent jelly-like substance within a **cell** – *i.e.* the **cytoplasm** (which contains various **organelles**) and the **nucleus**.

■ **Protozoa** Subkingdom or phylum of microscopic unicellular organisms which range from plant-like forms to types that feed and behave like animals. They have no common body shape (and some, such as amoeba, have no fixed shape at all) but all have specialized **organelles**. Their basic mode of reproduction is by **binary fission**, although multiple fission and conjugation occur in some species. Some protozoans are colonial and many are parasitic, inhabiting freshwater, marine and damp terrestrial environments. [2.5.b]

■ **protozoan** Member of the **Protozoa**. [2.5.b]

proventriculus *1.* Anterior glandular part of a bird's stomach, which secretes gastric juice. *2.* In insects and crustaceans, the **gizzard**.

pseudocarp Fruit that incorporates bracts, an inflorescence or receptacle in addition to the **ovary** of the flowering plant. Alternative name: false fruit.

pseudoparenchyma Fungal or algal tissue in which the filaments or **hyphae** are no longer discrete but have become an interwoven mass, falsely resembling **parenchyma**; *e.g.* stipe of a mushroom, thallus of a red alga.

■ **pseudopodium** Part of an **amoeba** or similar **protozoan** that is extruded from its unicellular body. Pseudopodia are used for locomotion and to engulf food particles (for digestion).

■ **pseudopregnancy** In some female mammals, physiological state resembling **pregnancy** without the formation of embryos. Alternative name: false pregnancy. [3.4.c]

■ **psychiatry** Study and treatment of disorders of the mind (*i.e.* mental disorders).

■ **psychology** Scientific study of the mind.

■ **Pteridophyta** Division of the plant kingdom that contains all vascular non-seed-bearing lower plants; *e.g.* clubmosses, ferns, horsetails. Pteridophytes are characterized by having a free-living **haploid gametophyte** generation which produces the male antheridia and the female archegonia. [2.5.b]

pteridophyte Plant that is a member of the **Pteridophyta**. [2.5.b]

■ **puberty** Stage in development when a child gradually changes into an adult, during which **sex hormones** are produced (by ovaries or testes) and **secondary sexual characteristics** appear. [3.4.c] [3.6.c]

■ **pulmonary** Concerning the lungs and breathing (or the respiratory cavities of molluscs). [3.4.a] [3.5.b] [3.5.e] [3.8.a]

■ **pulmonary artery** In mammals, a paired **artery** that carries deoxygenated blood from the right ventricle of the heart to the lungs. It is the only artery that carries deoxygenated blood. [3.4.a] [3.5.b] [3.5.e] [3.8.a]

■ **pulmonary vein** In mammals, a paired **vein** that carries oxygenated blood from the lungs to the left atrium of the heart. It is the only vein that carries oxygenated blood. [3.4.a] [3.5.b] [3.5.e] [3.8.a]

■ **pulse** *1.* In medicine, regular expansion of the wall of an artery caused by the blood-pressure waves that accompany heartbeats. *2.* In botany, a plant of the pea family; a legume. [3.5.b] [3.5.e]

■ **pupa** Inactive stage, characterized by a distinct body form, in the life-cycle of insects that undergo **metamorphosis**, during which a **larva** is transformed into an **imago**. Alternative name: chrysalis (especially in butterflies and moths). [2.3.c] [2.4.a]

■ **pupil** Hole in the iris of the **eye** (which appears as a black circle in the front of the eye); it allows light to enter and pass through the lens to the retina. [15.6.b]

purine $C_5H_4N_4$ Heterocyclic nitrogen-containing **base** from which the bases characteristic of **nucleotides** and **DNA** are derived; *e.g.* **adenine**, **guanine**. Other purine derivatives include caffeine and uric acid.

■ **putrefaction** Largely **anaerobic** decomposition of organic matter by microscopic organisms (*e.g.* bacteria, fungi, etc.) which results in the formation of incompletely oxidized products. [5.2.a] [7.7.b]

pyramid of numbers The numbers of animals at each **trophic level** decreases, as can be seen by representing them pictorially as a pyramid. [2.7.b]

pyranose Any of a group of **monosaccharide** sugars (hexoses) whose molecules have a six membered heterocyclic ring of five carbon atoms and one oxygen atom. *See also* **furanose**.

pyrenocarp Alternative name for **drupe**.

pyrenoid Spherical **protein** body found in the **chloroplasts** of some **algae** (*e.g.* Chlamydomonas).

pyridoxine Crystalline substance from which the active **coenzyme** forms of **vitamin** B_6 are derived. It is also utilized as a potent growth factor for **bacteria**.

pyrimidine $C_4H_4N_2$ Heterocyclic organic **base** from which bases found in **nucleotides** and **DNA** are derived; *e.g.* **uracil, thymine** and **cytosine**. Its derivatives also include barbituric acid and the barbiturate drugs.

pyrrole $(CH)_4NH$ Heterocyclic organic compound whose ring contains four carbon atoms and one nitrogen atom. A liquid **aromatic compound**, its derivatives are important biologically; *e.g.* haem, chlorophyll.

pyruvate Ester or **salt** of **pyruvic acid**.

pyruvic acid $CH_3COCOOH$ Simplest keto-acid, important in making energy available from ingested food. It is the product of the first stage of **respiration (glycolysis)**. If oxygen is available, the acid is broken down in the **Krebs cycle** (citric or tricarboxylic acid cycle) to yield energy. [3.6.a] [3.6.b]

Q

■ **quadrat** In an ecological survey of an area of ground, a small square (usually 1 square metre) within which all species are recorded or measured. [2.7.b]

■ **quadrate** One of a pair of bones in the upper jaw of fish, amphibians, reptiles and birds that has evolved into the incus (an **ear ossicle**) in mammals.

quantasome One of the tiny, semi-crystalline particles that occur in disc-shaped arrangements within the **chloroplasts** of plant cells, and which are believed to be the centre of light-processing during **photosynthesis**.

■ **quinine** Colourless crystalline **alkaloid**, obtained from the bark of the cinchona shrub, once much used in the treatment and prevention of malaria. [2.9.b]

R

■ **rabies** Virus disease, mainly affecting carnivores but which can afflict any warm-blooded animal, that is usually transmitted by bites from infected animals. Alternative (but erroneous) name: hydrophobia. [3.8.b]

race In biological classification, an alternative name for **subspecies**. *See also* **variety**.

racemic acid **Racemic mixture** of **tartaric acid**.

racemic mixture Optically inactive mixture that contains equal amounts of **dextrorotatory** and **laevorotatory** forms of an **optically active** compound.

racemization Transformation of **optically active** compounds into **racemic mixtures**. It can be effected by the action of heat or light, or by the use of chemical reagents.

racemose inflorescence Type of **inflorescence** in which the youngest flowers are at the growing tip of a flower stalk and the oldest ones are near the bottom.

■ **radial symmetry** Symmetry about any one of several lines or planes through the centre of an object or organism. *See also* **bilateral symmetry**.

■ **radiation sickness** Illness caused by exposure to excessive ionizing radiation. Alpha particles cannot penetrate skin and are not dangerous internally. Gamma radiation, X-rays and neutrons can penetrate the body and are thus the most harmful. [3.8.c]

radical In botany, relating to a root or stem base (*e.g.* radical leaves).

■ **radicle** Young **root** that arises in the **embryo** of plants.

■ **radioactive tracing** Use of radio-isotopes to study the movement and behaviour of an element through a biological or chemical system by observing the intensity of its **radioactivity**.

■ **radiobiology** Study of ionizing radiation in relation to living systems. It includes the effects of radiation on living organisms and the use of radio-isotopes in biological and medical work. *See also* **radiotherapy**. [3.8.c]

■ **radiocarbon dating** Method of estimating the ages of carbon-containing (*e.g.* wooden) archaeological and geological specimens that are up to 50,000 years old. A **radio-isotope** of carbon, carbon-14 (C-14), is present in **carbon dioxide** and becomes assimilated into plants during **photosynthesis** (and into animals that eat plants). The C-14 present in 'dead' carbonaceous materials decays and is not replaced (*see* **radioactive decay**). By comparing the **radioactivities** of the 'dead' and 'live' materials, the age of the former can be estimated, because the **half-life** of C-14 is known. A similar technique, potassium-argon dating, is used to determine the age of rocks. Alternative name: radioactive dating. [3.8.c]

radiodiagnosis Branch of medical **radiology** that is concerned with the use of **X-rays** or **radio-isotopes** in diagnosis.

radiograph Photographic image that results from uneven absorption by an object being subjected to penetrating **radiation**. An X-ray photograph is a common example.

■ **radiography** Photography using **X-rays** or **gamma rays**, particularly in medical applications.

Radiolaria Order of single-celled **plankton** animals, spherical in shape, that use **silicon** as a supporting structure rather than the more usual **calcium** compounds.

radiolarian ooze Sediment that contains a high proportion of **silicon** from the skeletons of **Radiolaria**, which covers the sea floor in the deepest equatorial waters of the Indian and Pacific Oceans.

radiology Study of **X-rays, gamma rays** and **radioactivity** (including **radio-isotopes**), especially as used in medical diagnosis and treatment.

radio-opaque Resistant to the penetrating effects of radiation, especially **X-rays**, often used to describe substances injected into the body before a **radiography** examination.

■ **radiotherapy** Treatment of disorders (*e.g.* **cancer**) by the use of **ionizing radiation** such as **X-rays** or radiation from **radio-isotopes**. [4.8.c]

■ **radius** One of two bones in the forearm or foreleg of a tetrapod vertebrate (the other is the **ulna**).

■ **radula** Organ resembling rows of small teeth, present in plant-eating **molluscs** (*e.g.* snails), used for feeding. Located in the **buccal cavity**, it possesses a serrated edge for rasping plant material.

raffinose $C_{18}H_{32}O_{16}$ Colourless crystalline **trisaccharide carbohydrate** that occurs in sugar beet, which hydrolyses to the **sugars** galactose, glucose and fructose.

Rajiformes Order of marine fishes comprising the rays, skates, etc., characterized by extremely flattened bodies, and which is most closely related to the sharks.

■ **ratite** Member of a group of flightless running birds that includes ostriches, rheas, emus and cassowaries.

■ **reafforestation** Alternative term for **afforestation**.

recapitulation theory Largely discredited belief based on external observations that the developmental stages undergone by an individual organism – *i.e.* from egg to adult – reflect various stages in the **evolution** of the organism's species.

■ **receptacle** Region of a plant that bears flower-parts; the top of a flower stem. It bears the **sepals, petals, stamens, gynaecium** or **carpels**. Its shape varies in different species. [3.4.d]

■ **receptor** Sensory **cell**, which may be part of a group that form a **sense organ** capable of detecting stimuli. When a receptor is stimulated (*e.g.* by temperature or light), it produces electrical or biochemical changes that are relayed to the **nervous system** for processing. [3.7.c]

■ **recessive** Describing a **gene** that is expressed in the **phenotype** (appearance of an organism) only when it is **homozygous** in a cell; *i.e.* there have to be two recessive genes for their effect to be apparent. The presence of a **dominant allele** masks the effect of a recessive gene; *i.e.* in a combination of a dominant gene and a recessive gene, the dominant gene manifests itself in the phenotype. [4.5.a] [4.6.a] [4.8.a]

■ **recipient** Organism that receives material from another, *e.g.* as in the taking up of **DNA** by one **bacterium** from another.

■ **reciprocal cross Hybrid** of two plants that results from mutual **pollination**.

recombinant DNA Type of **DNA** that has genes from different sources, genetically engineered using **recombination**.

■ **recombination** Process by which new combinations of characteristics not possessed by the parents are formed in the offspring. It results from **crossing over** during **meiosis** to form **gametes** that unite during fertilization to form a new individual. Genetic engineers have developed techniques for artificially recombining strands of **DNA** (to make recombinant DNA). [4.9.b] [4.10a]

■ **rectum** Final part of the **intestine**, through which **faeces** are passed and where they may be stored temporarily after reabsorption of water. [3.4.a] [3.5.e] [3.8.a]

■ **recycle** Use again. In nature, substances are recycled after some form of decomposition; *e.g.* nitrogen is released into the soil after plants and animals decay, to become available for other growing plants. [2.4.b] [2.6.b]

■ **red blood cell** Alternative name for an **erythrocyte**, also known as a red cell or red corpuscle.

■ **reducing agent** Substance that causes chemical **reduction**, often by adding hydrogen or removing oxygen; *e.g.* carbon, carbon monoxide, hydrogen.

reductase Enzyme that causes the **reduction** of an organic compound.

■ **reduction** Chemical reaction that involves the addition of hydrogen or removal of **oxygen** from a substance, often by the action of a **reducing agent**.

■ **reduction division** Alternative name for **meiosis**.

■ **reflex** In biology, sequence of nerve impulses that produces a faster involuntary response to an external stimulus than the corresponding voluntary response (which has to pass via the brain). [3.7.c]

■ **regeneration** Regrowth of tissue to replace that which has been damaged or lost (*e.g.* growth of a plant from a cutting, regeneration of a lost limb in starfish and crabs, and wound healing in higher mammals).

■ **relict** Species or population that has survived, often with a greatly reduced range, when all related organisms have become extinct. [2.4.c] [4.9.a]

renin Enzyme produced by the kidney that constricts arteries and thus raises blood pressure.

■ **rennin Enzyme** found in gastric juice that curdles milk. It is the active ingredient of rennet. [3.7.b]

replicase Enzyme that promotes the synthesis of **DNA** and **RNA** within living cells.

■ **reproduction** Procreation of an organism. **Sexual reproduction** involves the fusion of **sex cells** or **gametes** and the exchange of genetic material, thus bringing new vigour to a species. **Asexual reproduction** does not involve gametes, but usually the vegetative proliferation of an organism (*see* **vegetative propagation**). [3.2.a]

■ **reptile** Member of the animal class **Reptilia**. [2.3.b] [2.4.a] [2.5.b]

Reptilia Class of **poikilothermic** ('cold-blooded'), principally egg-laying vertebrates characterized by a body covering formed of scales. Reptiles include snakes, lizards, crocodiles, turtles, etc. [2.3.b] [2.4.a] [2.5.b]

■ **resin** Organic compound that is generally a viscous liquid or semi-liquid which gradually hardens when exposed to air, becoming an amorphous, brittle solid. Natural resins, found in plants, are yellowish in colour and insoluble in water, but are quite soluble in organic solvents. Many coniferous trees have an aromatic smell caused by resins.

■ **respiration** *1.* Release of energy by living organisms from the breakdown of organic compounds. In **aerobic respiration**, which occurs in most cells, **oxygen** is required and **carbon**

dioxide and water are produced. Energy production is coupled to a series of **oxidation-reduction reactions,** catalysed by **enzymes**. In **anaerobic respiration** (*e.g.* **fermentation**), food substances are only partly broken down, and thus less energy is released and oxygen is not required. *2.* Alternative name for breathing. [3.6.a]

■ **respiratory movement** Movement by an organism to allow the exchange of respiratory gases, *i.e.* the taking up of **oxygen** and release of **carbon dioxide**. In mammals such as human beings this entails breathing, involving movements of the chest and **diaphragm**. In fish, water is passed over the **gills** for gaseous exchange. [3.6.a]

■ **respiratory organ** Organ in which **respiration** (breathing) takes place. In mammals (*e.g.* human beings), the process is carried out in the **lung**; in fish, the **gills**. There gaseous exchange takes place (usually of **oxygen** and **carbon dioxide**). [3.6.a]

■ **respiratory pigment** Substance that can take up and carry **oxygen** in areas of high oxygen concentration, releasing it in parts of the organism with low oxygen concentrations where it is consumed, *i.e.* by **respiration** in cells. In vertebrates the respiratory pigment is **haemoglobin**; in some invertebrates it is **haemocyanin**. [3.6.a]

■ **respiratory quotient** (RQ) Ratio of **carbon dioxide** produced by an organism to the **oxygen** consumed in a given time. It gives information about the type of food being oxidized. *E.g.* **carbohydrate** has an RQ of approximately 1, but if the RQ becomes high (*i.e.* little oxygen is available), **anaerobic respiration** may occur. [3.6.a]

■ **response** Physical, chemical or behavioural change in an organism initiated by a **stimulus**. [3.7.c]

resting potential **Potential difference** between the inner and outer surfaces of a resting **nerve**, which is about −60 to −80 mV. It happens when the nerve is not conducting any impulse, and is in contrast to the **action potential**, which occurs during the application of a stimulus and brings about a rise in the potential difference to a positive value.

restriction enzyme **Enzyme** (a **nuclease**) produced by some **bacteria** that is capable of breaking down foreign **DNA**. It cleaves double-stranded DNA at a specific sequence of **bases**, and the DNA of the bacteria is modified for protection against degradation. Restriction enzymes are used widely as tools in **genetic engineering** for cutting DNA. Alternative name: restriction endonuclease.

■ **reticulum** Second chamber of the stomach of a **ruminant**. *See also* **endoplasmic reticulum**. [3.5.a]

■ **retina** Light-sensitive tissue at the back of the vertebrate **eye**, made up of a network of interconnected nerves. The first cells in the network are photo-receptors consisting of **cones** (which are sensitive to colour) or **rods** (which are sensitive to light). They act by means of visual pigments (*e.g.* **rhodopsin**) which cause impulses to be transmitted to the visual centre of the brain via the **optic nerve**. [3.4.a] [3.5.e] [3.8.a] [15.6.b]

retinol Fat-soluble **vitamin** found in plants, in which it is formed from **carotene**. Alternative name: vitamin A.

retrix One of a bird's tail feathers; most species have 12 retrices.

retrovirus Member of a group of **viruses** that contain **RNA** as their genetic material. They use an RNA-dependent **DNA polymerase** or reverse transcriptase **enzyme** to carry out **transcription**. Many RNA viruses are **carcinogenic** in their hosts, which include mammals. Alternative name: oncovirus.

rhabdovirus Member of a group of viruses that can infect multi-cellular animals and plants. One type causes **rabies**.

rhinovirus Member of a group of viruses that infect the respiratory tract of vertebrates and which are one of the main causative agents of the common cold.

■ **rhizoid** Small **root**-like structure, composed of one or few cells, used for anchorage by some lower plants (*e.g.* liverworts, mosses). [3.4.a] [3.5.e] [3.7.a] [3.8.a]

■ **rhizome** Underground main **stem** used for food storage by some plants (*e.g.* many grasses, iris). [3.4.a] [3.5.e] [3.7.a] [3.8.a]

■ **rhodopsin** Protein (derived from vitamin A) in the **rods** of the **retina** of the eye which acts as a light-sensitive pigment; the action of light brings about a chemical change that results in the production of a nerve impulse. Alternative name: visual purple.

■ **riboflavin** Orange water-soluble crystalline solid, member of the **vitamin** B complex. It plays an important role in growth. Alternative names: riboflavine, lactoflavin, vitamin B_2.

■ **ribonucleic acid** *See* **RNA**. [4.9.d]

■ **ribose** $C_5H_{10}O_5$ **Optically active pentose** sugar, a component of the **nucleotides** of **RNA** (ribonucleic acid).

■ **ribosome** Particle present in the **cytoplasm** of cells, often attached to the **endoplasmic reticulum**, that is essential in the biosynthesis of **proteins**. Ribosomes are composed of protein and **RNA**, and are the site of attachment for **messenger RNA** during protein synthesis. They may be associated in chains called polyribosomes. [4.9.d]

■ **rickets** Disorder that results from a deficiency of **vitamin** D.

It mainly affects children and can cause deformed limbs. [3.8.a] [3.8.b] [3.8.c]

rickettsiae Group of micro-organisms, often classified as being part-way between **bacteria** and **viruses**, that are parasitic on the cells of arthropods (lice, mites and ticks) and vertebrates. Some can cause serious disorders (*e.g.* typhus in human beings).

■ **Ringer's fluid/solution** Physiological **saline** solution used for keeping **tissues** and **organs** alive outside the body (in vitro). It is similar in composition to the fluid that naturally bathes cells and tissues, maintaining a constant internal environment. It contains chlorides of sodium, potassium and calcium. It was named after the British physiologist Sydney Ringer (1835–1910).

■ **ringworm** *See* **tinea**.

■ **RNA** Abbreviation of ribonucleic acid, one of the **nucleic acids** present in cells, the other being **DNA**. It is composed of **nucleotides** that contain **ribose** as the sugar. RNA contains the bases **adenine, guanine, cytosine** and **uracil. Messenger RNA** takes part in **transcription** or copying of the **genetic code** from a DNA template. **Transfer RNA** and **ribosomal RNA** take part in **translation** or **protein synthesis**, all of which occur in **prokaryotes** and **eukaryotes**.

■ **RNA virus** A **virus** that has **RNA** as its genetic material (instead of **DNA**).

■ **rod** Type of sensory cell present in the **retina** of the vertebrate **eye**. It is stimulated by light and is concerned with vision in low illumination. The absorption of light energy (photons) by the visual pigment **rhodopsin** present in the rod causes a nervous impulse, which travels along the **optic nerve** to the brain. *See also* **cone**. [15.6.b]

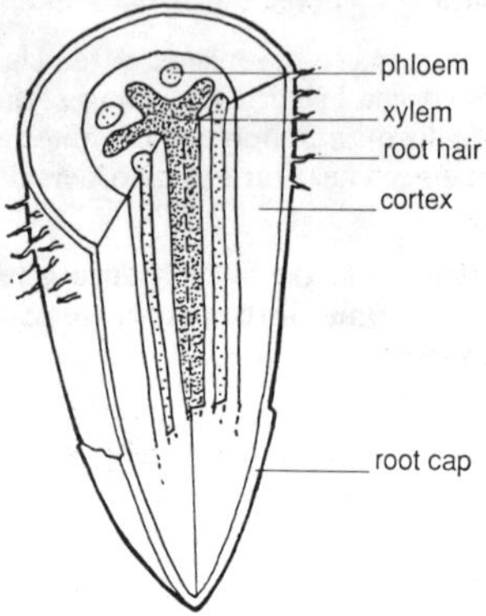

Structure of a root tip

■ **root** Structure in **vascular plants** whose function is anchorage and the uptake of water and **mineral salts** by **osmosis**. Roots are usually partly or completely underground. The vascular tissues form a central core, unlike those in a **stem**.

root hair Single cell in the **outer layer** of a young root which has a very large surface area in relation to its volume, thus increasing its efficiency for absorbing water.

root nodule Lump that occurs on the roots of leguminous plants (*e.g.* peas, beans, clover) which contains bacteria that can bring about the **fixation of nitrogen**.

root pressure One of the ways in which water rises up a plant (the others being due to upward 'pull' of **capillarity** and **transpiration**).

■ **rostrum** *1.* Forward projection of a fish's head beyond its mouth. *2.* Specialized piercing and sucking mouthparts of a **bug**.

Rotifera Phylum of small (mostly less than 2 mm in length) aquatic animals, which swim by beating a 'wheel' of cilia. Most rotifers are found in fresh water. Alternative name: wheel-animalcules.

■ **roughage Fibre** that forms the bulking agent in the human diet and which is essential for the proper working and health of the alimentary canal.

■ **round window** Lower of two membranous areas on the **cochlea** of the inner **ear** (the other is the **oval window**). Alternative name: fenestra rotunda. [3.4.a] [3.5.e] [3.8.a] [14.6.b]

r-selection Survival strategy that is characteristic of colonizing species, and which tends towards high birth rates combined with a short lifespan for individuals. *See also* **k-selection**.

■ **rubber** Elastic substance obtained from plant **latex**. It is a **polymer**, containing long chains of the monomer **isoprene**. Alternative name: natural rubber.

■ **ruminant** Animal of the suborder Ruminantia, which includes cattle, deer and other mammals that chew the cud. Ruminants are characterized by complex digestive systems, with multi-chambered stomachs, that can break down **cellulose** with the aid of bacterial **fermentation**. [2.5.d] [2.10.a]

S

saccharide Simplest type of **carbohydrate,** with the general formula $(C_6H_{12}O_6)_n$, common to many **sugars**. *See also* **disaccharide**; **monosaccharide**.

saccharimetry Measurement of the concentration of **sugar** in a solution from its **optical activity**, by using a polarimeter.

■ **saccharin** $C_6H_4SO_2CONH$ White crystalline organic compound that is about 550 times sweeter than **sugar**; an artificial sweetener. It is almost insoluble in water and hence it is used in the form of its soluble sodium salt. Alternative names: 2-sulphobenzimide, saccharine.

saccharose Alternative name for **saccharide**.

■ **saline** Salty; describing a solution of **sodium chloride** (common salt).

■ **saliva** Neutral or slightly alkaline fluid secreted by the salivary glands in the mouth. It lubricates food during chewing and aids digestion. It consists of a mixture of **mucus** and the **enzyme amylase** (ptyalin), which breaks down **starch** to **maltose**. [3.5.c] [3.7.b]

■ **Salmonella** Genus of anaerobic bacteria that can cause disorders in humans, including enteric fever, paratyphoid, typhoid fever and a common type of food poisoning.

■ **salpingectomy** Method of sterilizing a female by surgically cutting (and tying the cut ends) of the oviducts, or Fallopian tubes, thus making it impossible for an ovum (egg) to pass from an ovary to the uterus (womb).

■ **salt** *1*. Product obtained when a **hydrogen** atom in an **acid** is replaced by a **metal** or its equivalent (*e.g.* the ammonium ion NH_4^+). It results from the reaction between an acid and a base. *2.* Common salt, **sodium chloride**. [3.5.b]

samara Winged seed (*e.g.* those produced by sycamore and maple trees).

■ **saprophyte** Organism that feeds on dead organic matter (*e.g.* many **bacteria** and **fungi**). Saprophytic activity is the first step in the decomposition of dead animals and plants, and consequently is important in the recycling of elements. Alternative name: saprotroph. [5.4.a]

saprotroph Alternative name for **saprophyte**.

satellite DNA Fraction of **DNA** with significantly different **density** and thus **base** composition from most of the DNA in an organism.

■ **scapula** Alternative name for the shoulder blade. [3.3.a] [3.5.e]

schistosomiasis Alternative name for **bilharzia**.

schizogony Form of **asexual reproduction** employed by some single-celled animals, in which a parent cell divides into more than two independent cells. *See also* **binary fission**.

Schwann cell Cell that produces the **myelin sheath** that surrounds a nerve cell (**neurone**). Schwann cells are in close contact with the **axon** of the neurone and are separated by gaps called **nodes of Ranvier**. They were named after the German physiologist Theodor Schwann (1810–82).

sciaphilic Describing a plant that grows only in shady conditions.

■ **scion** Piece of a plant, usually a shoot, that is inserted into the body of another plant when making a **graft**. [3.9.b]

■ **sclerenchyma** Simple plant tissue composed of cells with thickened walls containing **lignin**, used to give support. The tissue may be composed of fibres or rounded cells.

scleroprotein Member of a group of fibrous **proteins** that provide organisms with structural materials (*e.g.* **collagen**, **keratin**).

■ **sclerotic** Outermost of the three layers that form the eyeball (outside the **choroid** and **retina**). [15.6.b]

■ **scrotum** Sac present in males of some mammals that contains the **testes**, positioned outside the body cavity so that their temperature is cool enough for **sperm** production. [3.4.c]

■ **scurvy** Disorder of the skin and gums caused by lack of vitamin C (ascorbic acid) in the diet, once common among sailors because, on long voyages, they had little or no fresh fruit or vegetables in their diet. [3.8.b]

■ **seaweed** Type of **alga** whose **habitat** is seawater, divided into red, brown and green seaweeds according to which **pigment** is present.

■ **sebaceous gland** Small gland found in large numbers in the skin of mammals, usually alongside a hair follicle, that secretes the protective skin oil **sebum**.

■ **sebum** Waxy material secreted by **sebaceous glands**, which helps to keep skin waterproof.

■ **secondary growth** Growth exhibited by some woody plants which takes place in the **stem** and **root**, increasing their girth. It occurs as a result of division of **cambium** located between the **xylem** and **phloem**. Alternative name: secondary thickening.

■ **secondary sexual characteristics** Features that develop in some animals after the onset of **puberty**, distinguishing males from females, but not required for sexual function. They result from the actions of **sex hormones**, principally **testosterone** and **oestrogen**. [3.4.c]

■ **secondary thickening** Alternative name for **secondary growth**.

■ **secretion** Release of a substance by a cell or **gland** with a specialized function, *e.g.* secretion of digestive **enzymes** by cells of the small intestine, or secretion of **hormones** by the pituitary.

■ **sedative** Drug that calms without (in normal doses) causing loss of awareness or consciousness. In larger doses, some sedatives become sleep-inducing drugs. *See also* **tranquillizer**. [3.4.b] [3.8.b]

■ **sedge** Member of a large group of grasslike plants that have **rhizomes**, and which are found throughout the world, particularly in marshy conditions. Some species are important as animal fodder. [2.1.a] [2.5.a]

■ **sedimentation** Removal of solid particles from a **suspension** by gravitational force or in a **centrifuge**.

■ **seed** Structure formed from the **ovule** of a plant after **fertilization**. It contains the developing **embryo** and is a highly resistant structure which can withstand adverse conditions. Dispersal of seeds may be by wind, water or animals and is important in the spreading and colonization of plants. [2.4.c] [2.6.a] [3.4.a] [3.4.d]

■ **segment** One of several adjacent parts of an organism's body that have similar or identical forms (*e.g.* in some worms and insects).

■ **segregation** Separation of a pair of **alleles** in a **diploid** organism during **meiosis** in the formation of **gametes**. A gamete receives one of the two alleles in a diploid organism because it receives only one of a pair of **homologous chromosomes**. [4.5.a]

selection *See* **artificial selection; natural selection.**

selective breeding *See* **artificial selection.**

■ **selectively permeable membrane** Alternative name for **semi-permeable membrane**.

self-fertilization *See* **self-pollination.**

■ **self-pollination** Type of **fertilization** in plants in which pollen is transferred from the anther to the stigma of the same flower. Many plants have a mechanism which ensures that, if cross-pollination (between two different plants of the same species) does not occur, self-pollination does. [3.4.d]

■ **semen** Fluid produced in male reproductive organs of many animals. It contains **sperm**, and in mammals secretions from the accessory sex glands. [3.4.c]

■ **semicircular canal** Part of the **ear** that is involved in maintaining balance. *See* **labyrinth**. [3.4.a] [3.5.e] [3.8.a] [14.6.b]

■ **seminal vesicle** Organ in the **testes** that is used for storing **sperm**. [3.4.c]

■ **seminiferous tubule** One of many tubes within the **testes** in which **sperm** are made.

■ **semi-permeable membrane** Porous **membrane** that permits the passage of some substances but not others; *e.g.* **plasma membrane**, which permits entry of small molecules such as water but not large molecules, allowing **osmosis** to occur. Such membranes are extremely important in biological systems and are used in **dialysis**. Alternative name: selectively permeable membrane. [3.8.a]

■ **senescence** Processes that mark the final stages of an organism's natural life-span. In plants senescence is usually associated with flowering and fruiting.

■ **sense organ** Group of **receptors** specialized to react to (detect) a certain **stimulus** (*e.g.* the **eye** to light, the **ear** to sound, and **chemoreceptors** in the tongue and nose to tastes and smells). [3.7.c]

■ **senses** The five primary senses, common to most vertebrates but sometimes lacking in less highly evolved animals, are sight, hearing, taste, smell and touch, to which may be added the sense of balance. They are effected by various **sense organs**. [3.4.a] [3.7.c]

■ **sepal** Part of certain flowers that forms the **calyx**. [3.4.a] [3.5.e] [3.8.a]

■ **septicaemia** Disorder that results from the presence of bacteria, or their toxins, in the bloodstream. Alternative name: blood poisoning.

■ **septum** Dividing wall found in biological systems, *e.g.* between the nostrils or between the two halves of the heart.

sere Any of the characteristic **communities** that occur in sequence during the process of plant **succession**.

serine $CH_2OHCHNH_2COOH$ White crystalline **amino acid**, present in many **proteins**. Alternative name: 2-amino-3-hydroxypropanoic acid.

■ **serology** Branch of **immunology** concerned with reactions between **antibodies** of one organism with **antigens** of the **serum** of another.

serotinal Describing biological activity or events that take place during late summer.

serotonin Substance derived from **tryptophan** and found in blood **serum**, used as a **neurotransmitter** and **vasoconstrictor**. Alternative names: 5H, 5-hydroxytryptamine.

■ **serum** Constituent of **plasma** of blood, which contains all the substances in plasma except for **fibrinogen**. [3.8.b]

■ **sessile** *1.* Not having a stalk (of leaves). *2.* Non-mobile, describing particularly animals that permanently anchor themselves to a surface such as the sea bed.

■ **sewage** Faeces, urine, washing water and surface water from homes and factories. These waste materials pass along sewers to works where they are treated to make them harmless. Any useful materials are extracted and recycled; treated water is returned to rivers or the sea.

■ **sex cell** Alternative name for **gamete**.

■ **sex chromosome Chromosome** that carries the **genes** determining sex. In mammals the female possesses two identical sex chromosomes or X- chromosomes, whereas in the male the two sex chromosomes differ, one being an X- and the other a Y-chromosome. *See also* **heterogametic**; **homogametic**; **sex determination**. [4.7.a]

■ **sex determination** Inheritance of particular combination of **sex chromosomes**, which is the deciding factor in whether an organism is male or female. Inheritance of a **homologous** pair of sex chromosomes predisposes the organism to one sex (*e.g.* in mammals, the female). Inheritance of a pair of dissimilar sex chromosomes determines the other sex (in mammals, the male). *See also* **heterogametic**; **homogametic**. [4.7.a]

■ **sex hormone Hormone** that determines **secondary sexual characteristics** and regulates the reproductive behaviour of an organism. In mammals, the sexual cycle of the female is controlled by such hormones (*see* **oestrous cycle**). In males the **gonads** are regulated. [3.8.a] [3.9.b] [4.7.a]

■ **sex linkage** Distribution of **genes** according to the sex of an organism because they are carried on the **sex chromosomes**. In human males a **recessive** gene carried on the **X-chromosome** will be expressed because no corresponding **allele** is present on the **Y-chromosome** to mask it, the Y-chromosome being shorter than the X-chromosome. In the female the corresponding allele will be present on the other X-chromosome, and for this reason human males have a predisposition to recessive sex-linked disorders (*e.g.* haemophilia, colour-blindness), whereas human females are more often **carriers** rather than sufferers of such disorders. [4.7.a]

■ **sex ratio** Ratio of the number of males to the number of females in a population. It may be expressed as the number of males to every 100 females. [4.7.a]

■ **sexual reproduction Reproduction** of an organism that involves the fusion of specialized sex cells or **gametes** (which are **haploid**) to form **diploid** progeny. It is important in bringing new vigour to a **species** by the mixing of genetic material from the parents to give a genetically different organism. *See also* **asexual reproduction**; **egg**; **fertilization**; **sperm**. [3.4.c] [4.7.a]

■ **shell** *1.* Protective chalky or leathery outer covering of an egg (*e.g.* a bird's or reptile's egg). *2.* Hard chalky outer covering of some types of **mollusc** (*e.g.* mussel, snail), secreted by the **mantle**. *3.* **Exoskeleton** of an **arthropod** (*e.g.* barnacle, crab, lobster). *4.* **Carapace** of a tortoise or turtle. *5.* Hard outer layer of some fruits (*e.g.* nuts).

■ **short-sightedness** Alternative name for **myopia**. [15.6.c]

■ **shoulder blade** Alternative name for **scapula**.

■ **shunt** *1.* In medicine, a surgically implanted tube or vessel that diverts the flow of fluid (*e.g.* to by-pass an obstruction or to drain an area). *2.* Small blood vessel between capillaries that normally carry blood to and from the surface of the skin. When the skin is cold, blood is diverted along the shunt vessels to retain body heat (thus causing very pale, white skin). [3.8.a]

■ **sickle-cell anaemia** Inherited type of **anaemia** that generally affects only black people. An abnormality of the **haemoglobin** in the blood causes red blood cells to take the shape of crescents, which cannot easily pass along narrow blood vessels and tend to be broken down instead (sometimes leading to **thrombosis**). [4.6.b] [4.8.b]

■ **sieve tube** Tubular element that makes up the **phloem** in a vascular plant, through which **translocation** occurs. It is composed of cellulose-walled sieve elements which are joined via pores, and have lost their **organelles**.

■ **sinus** Irregular cavity or depression that forms part of an animal's anatomy; *e.g.* sinuses in the bones of the face in mammals.

siphon In biology, *1.* a sucking mouthpart (*e.g.* as in fleas); *2.* a tubular organ through which liquid (often water) passes into or out of an animal's body (*e.g.* as in some molluscs). Alternative name: syphon.

■ **skeleton** Structure that supports the **tissues** and **organs** of an animal and is attached to **muscles** and **ligaments** to allow locomotion. An **endoskeleton** is internal, made of **bone** or **cartilage**, and possessed by **vertebrates**. **Exoskeletons** lie outside the muscles, *e.g.* in **arthropods**. Some **invertebrates** possess a hydrostatic skeleton (water vascular system) which consists of fluid under pressure. [3.3.a]

■ **skin** Organ in mammals that protects the body from invasion by **pathogens** and prevents water from entering. In warm-blooded animals the skin also takes part in temperature regulation, *e.g.* through sweating and the constriction and dilation of its **blood vessels**. It consists of **epithelial tissue** and **connective tissue** arranged in two major layers, the thin outer epidermis and the thicker underlying dermis. [3.8.b]

■ **skull** Bones that form the head and face, including the **cranium** and **jaws**.

■ **small intestine** Part of the digestive tract in mammals which is composed of the **duodenum** and **ileum**, and is the main site of **digestion** and absorption in the gut. **Bile**, pancreatic juice and intestinal juice are liberated in it, to supply many digestive **enzymes**. *See also* **intestine**. [3.5.c] [3.7.b]

■ **smell** One of the primary **senses** that enables animals to detect odours, using **chemoreceptors** usually located in mammals in olfactory glands in the nasal cavity. [3.1.a] [3.4.a] [3.5.e]

■ **smooth muscle** Type of muscle in internal organs and tissues, not under voluntary control. Alternative name: involuntary muscle. [3.5.e] [3.8.a]

■ **sodium pump** Process by which potassium and sodium **ions** are transported across **membranes** that surround animal cells. [3.7.c]

■ **soft palate** Rear part of the roof of the mouth, consisting of muscle tissue covered by mucous membrane. The uvula hangs from the back of the soft palate.

solenocyte Alternative name for **flame cell**.

■ **somatic** Of the body. Somatic cells include all cells of an organism except for the **gametes** or sex-cells.

somatotrophin Alternative name for **growth hormone**.

sorbitol Alcohol formed by the **reduction** of **glucose**, used as a sweetening agent.

soredium Structure formed during the **asexual reproduction** of some **lichens** that contains both fungal and algal cells.

■ **species** Smallest group commonly used in biological **classification** and into which a genus is divided. Species are sometimes further divided into subspecies (*e.g.* races, varieties). Generally, no more than one type of organism is present in one species. Members of a species may breed with one another, but usually cannot breed with members of another species. Rarely, very closely related species interbreed to produce a **hybrid**. *See also* **binomial nomenclature**. [2.3.b] [2.4.a] [2.5.b]

■ **sperm** Abbreviation of **spermatozoon**, the **gamete** (sex-cell) produced by the male in many species of animals (in mammals, in the **testes**). It consists of a head containing genetic material in the **nucleus** (which is **haploid**), and usually possesses **cilia** or a **flagellum** for movement. [3.4.c]

■ **spermaceti** Waxy oil that occurs in tissues located behind the forehead of sperm whales. Its function is uncertain, but it is believed to be involved in echo-location. It can be used in the manufacture of medicines and cosmetics.

■ **spermatocyte** Cell from which **sperm** (spermatozoa) are derived through **spermatogenesis**. Primary spermatocytes are **diploid**; after **meiosis,** secondary spermatocytes which are **haploid** are formed. These further divide to produce spermatids, which differentiate to form sperm. [3.4.c]

■ **spermatogenesis** Formation of **sperm** in the **testis**. It commences with the repeated **mitosis** of primordial germ cells to form spermatogonia, which grow to form a primary **spermatocyte**. This divides by meiotic (reduction) division to produce four **haploid** spermatozoa. [3.4.c]

■ **Spermatophyta** Division of the plant kingdom that contains all seed-bearing plants. Spermatophytes may be divided into **angiosperms** (conifers, etc.) and **gymnosperms** (flowering plants) which can be further subdivided into **dicotyledons** (class Magnoliopsida) and **monocotyledons** (Liliopsida)).

■ **spermatophyte** Member of the plant division **Spermatophyta**.

■ **spermatozoon** *See* **sperm**.

■ **sphincter** Circular muscle that controls the flow of a liquid or semi-solid through an orifice (*e.g.* the anal sphincter, round the anus).

■ **sphygmomanometer** Instrument for measuring blood pressure.

■ **spinal column** *See* **spine**.

■ **spinal cord** Part of the **central nervous system** (CNS) in vertebrates that is enclosed within the spine. It consists of a hollow nerve tube containing many interconnecting **neurones** and connected to the **spinal nerves**.

■ **spinal nerve** Any of several peripheral nerves arising from the **spinal cord** which are connected to **receptors** and **effectors** in other parts of the body.

■ **spindle** Structure composed of **protein** fibres that is formed in the **cytoplasm** during **cell division**. It is used for the attachment of **chromosomes** and thought to assist in their movement to the poles of the cell. *See also* **meiosis**; **mitosis**.

■ **spine** Backbone; dorsally situated bony column composed of **vertebrae**, which enclose the **spinal cord**. Alternative name: spinal column. [3.3.a]

■ **spiracle** *1.* In cartilaginous fish, one of a pair of openings on the head near the gills. *2.* In insects, an opening that leads into a system of **tracheae**, used for **respiration**.

■ **spirochaete** Member of a group of **bacteria** characterized by a spiral shape, some of which are parasitic and cause disorders (*e.g.* syphilis, yaws, infectious jaundice).

■ **spirogyra** Commonest filamentous **alga** (genus *Spirogyra*) found in freshwater ponds, which has one or more spiral-shaped **chloroplasts** in each cell. [2.4.a]

splanchnology Branch of biology and medicine concerned with the organs within the central body cavity of vertebrates.

■ **spleen** Organ present in the abdomen of some vertebrates that aids in defence against invading organisms. It produces **lymphocytes** and also stores and removes red blood cells (erythrocytes) from the blood system. [3.4.a] [3.5.e] [3.8.a]

■ **sponge** Member of the animal phylum **Porifera**.

■ **spontaneous generation** Theory (now disproved) that living matter can arise from non-living matter. *See also* **abiogenesis; biogenesis**. [4.9.a]

■ **sporangium** Plant structure that produces **spores**.

■ **spore** *1.* **Haploid** reproductive structure formed in the **sporophyte** of plants. It can be widely dispersed by wind etc. and germinates to form a **gametophyte**. *2.* Resting form of an organism that is highly resistant to adverse conditions, *e.g.* in **bacteria**.

■ **sporophyll** Plant part that bears **sporangia**; *e.g.* leaves of ferns, cones of pteridophytes.

■ **sporophyte** Stage in the life cycle of a plant that is **diploid** but produces **haploid spores** which generate the **gametophyte**. *See also* **alternation of generations**.

■ **sport** In biology, an organism that possesses abnormal characteristics as a result of a naturally occurring **mutation**. The term is most often applied to plants. [4.8.c] [4.9.a]

■ **stamen** Part of a flower that bears the male reproductive structures. It consists of an **anther** in which the **pollen** grains develop, which in turn bear the male **gametes**. [3.4.d]

■ **stapes** One of the **ear** ossicles. Alternative name: stirrup. [3.4.a] [3.5.e] [3.8.a] [14.6.b]

■ **Staphylococcus** Type of Gram-positive **bacterium** characterized by its shape, which takes the form of irregular clusters (resembling a miniature bunch of grapes). Staphylococci cause various inflammatory disorders (*e.g.* boils, impetigo, osteomyelitis). *See also* **coccus**. [3.8.b]

■ **starch** $(C_6H_{10}O_5)_n$ Complex **polysaccharide carbohydrate**, a **polymer** of **glucose**, that occurs in all green plants, where it serves as a reserve energy material. It forms glucose on complete **hydrolysis**. Alternative name: amylum. [3.6.b] [3.7.b]

stasigenesis Process of non-evolution in which the characteristics of a species remain completely unchanged over very long periods of time, and the species becomes a 'living fossil' (*e.g.* coelacanth, tuatara).

■ **statocyst** Sensory organ found in some aquatic invertebrates (*e.g.* certain **crustaceans**) that consists of a cavity containing grains of a hard substance, and which is used to monitor changes in the animals' direction of movement.

■ **stearic acid** $CH_3(CH_2)_{16}COOH$ Long-chain **fatty acid** (a carboxylic acid) that occurs in most **fats and oils**. Its sodium and potassium **salts** are constituents of soaps. Alternative name: octadecanoic acid.

stele Vascular parts (**phloem** and **xylem**) of the roots and stems of a vascular plant.

■ **stem** Axis of a plant which is usually above ground, bearing leaves and flowers in vascular plants. Its arrangement of vascular tissues differs from that of a **root**. *See also* **corm; phloem; rhizome; secondary growth; tuber; xylem**. [3.1.a] [3.4.a] [3.5.e] [3.7.a] [3.8.a]

■ **sterilization** *1.* Treatment of an apparatus or substance (*e.g.* food) so that it contains no micro-organisms that could cause disease or spoilage, usually by means of high temperatures, gamma radiation, etc. *2.* Surgical treatment of an animal so that it cannot have offspring (*e.g.* in mammals by cutting or tying the Fallopian tubes of the female or vasa deferentia of the male; removing the ovaries or testes is a more drastic way of achieving the same effect).

■ **sternum** Bone in tetrapod vertebrates on the ventral side of the **thorax**, parallel to the **spine**, to which most of the ribs are attached. Alternative name: breastbone.

■ **steroid** Any of a group of naturally occurring organic compounds, widely found in animal tissues. Most have very important physiological activities (*e.g.* adrenal hormones, bile acids, sex hormones, sterols). Some can be made synthetically (*e.g.* for use as contraceptive pills and for treatment of hay fever). *See also* **anabolic steroid**.

■ **sterol** Subgroup of **steroids** or steroid alcohols. They include **cholesterol**, abundant in animal tissues, which is the precursor of many other steroids.

■ **stethoscope** Instrument for listening to sounds made by internal organs (*e.g.* heart, lungs).

■ **stigma** *1.* In a flower, swelling at the apex of the **style** in the

carpel onto which **pollen** is transferred. *2.* In some **flagellates**, pigment spot sensitive to changes in light. [3.4.a] [3.5.e] [3.8.a]

■ **stimulant** Any **drug** that produces a stimulating effect; *e.g.* alcohol and nicotine in very small quantities (larger quantities are depressants), amphetamines, caffeine. Many stimulants can be addictive (*see* **addiction**). [3.6.e]

■ **stimulus** Environmental factor that is detected by a **receptor** and induces a **response** from an **effector**.

stolon *1.* In botany, stem that grows parallel with the ground; a runner (*see* **vegetative propagation**). *2.* In zoology, stalk-like body part that anchors many types of aquatic invertebrates and which bears polyps.

■ **stoma** Specialized pore found on the stems and leaves (many on the undersurfaces) of vascular plants which allows the exchange of gases and loss of water during **transpiration**. When the stomatal guard cells are turgid, the stomata are open; when the guard cells are flaccid, they are closed. [3.4.a] [3.5.e] [3.6.b] [3.8.a]

■ **stomach** Muscular sac present in vertebrates in which food is partly digested and stored after passage through **oesophagus** (gullet). Hydrochloric acid is secreted in the stomach, as is the **enzyme pepsin**, which begins the digestion of **proteins**. [3.5.c] [3.7.b]

■ **stomata** Plural of **stoma**.

■ **stomatal guard cell** One of a pair of sausage-shaped cells within the thickness of cellulose in the cell wall of a leaf. When the guard cells are turgid they curve, and the pore between them (**stoma**) opens. [3.6.b] [3.8.a]

■ **Streptococcus** Type of Gram-positive **bacterium** characterized by its shape, which takes the form of a chain. Streptococci cause various disorders (*e.g.* erysipelas, scarlet fever).

■ **streptomycin** **Antibiotic** that works by inhibiting **protein synthesis** in bacterial cells.

■ **stress** In biology, any environmental factor, or combination of factors, that has adverse effects on the structure or behaviour of an organism.

■ **striated muscle** **Muscle** that contains well-aligned threads of **protein**, which enable it to contract strongly in a particular direction. Alternative name: voluntary muscle. [3.5.e]

■ **stridulation** Process by which many insects (*e.g.* grasshoppers, crickets) produce sound by rubbing body parts together. Typically, a toothed 'file' on a wing or leg is moved rapidly against the hardened edge of another wing.

strigil Cluster of hairs on the forelegs of some insects (*e.g.* butterflies, wasps, bees) that is used for grooming.

strobilus Alternative name for a (plant) **cone**.

■ **stroke** Sudden loss of part of the brain's function due to lack of oxygen caused by the blocking of a blood vessel serving the brain by a clot (**thrombosis**) or by a vessel bursting (cerebral haemorrhage). Alternative name: apoplexy.

stroma Matrix in which the **lamellae** of a green plant's **chloroplasts** are embedded, and the site of dark reactions of **photosynthesis** (which build up **carbohydrates**).

■ **strychnine** White crystalline insoluble **alkaloid** with a bitter taste, one of the most powerful **poisons** known. Alternative name: vauqueline.

■ **style** Part of the **carpel** of a flowering **plant** that bears the **stigma**. [3.5.d]

■ **subcutaneous tissue** Layer of **tissue** below the **dermis** of the **skin** in vertebrates. It often contains deposits of **fat**.

sub-imago Pre-adult winged insect form that is unique to the mayflies (order Ephemeroptera).

■ **subspecies** Group of organisms within a **species** that have certain characteristics not possessed by other members of the species. Breeding may occur between members of different subspecies. *See also* **race; variety**. [2.5.b]

■ **substrate** *1*. Molecule on which an **enzyme** exerts its catalytic action. *2*. Substance upon which an organism grows or to which it is attached. [3.5.c]

■ **succession** Progressive change in the structure of a community when colonizing a **habitat** until a stable **climax community** is established. [2.9.a]

■ **succulent** Plant with swollen leaves or stems, adapted for living in dry habitats or similar conditions in which there is little fresh water (*e.g.* a salt marsh). *See also* **halophyte**; **xerophyte**.

■ **sucrase Enzyme** that breaks down **sucrose** into simpler sugars. Alternative name: invertase.

■ **sucrose** $C_{12}H_{22}O_{11}$ White soluble crystalline disaccharide which is obtained from sugar cane and sugar-beet; ordinary sugar, used to sweeten food. It is hydrolysed to **fructose** and **glucose**. Alternative names: cane-sugar, beet-sugar, sugar. [3.7.b]

■ **sugar** *1*. Crystalline soluble **carbohydrate** with a sweet taste; usually a **monosaccharide** or **disaccharide**. *2*. Common name for **sucrose**. [3.7.b]

■ **suspension** Mixture of insoluble small solid particles and a fluid through which the insoluble substance stays evenly distributed (because of molecular collisions and the fluid's viscosity, which prevents precipitation). Alternative name: suspensoid.

■ **suspensory ligament** One of the structures that hold the lens of the **eye** in position. [15.6.b]

■ **sweat** Watery fluid containing salts secreted from glands in the **skin**. **Evaporation** of sweat aids in cooling the body. Alternative name: perspiration. [3.8.a]

■ **swim bladder** Structure present in bony fish, used for controlling buoyancy by filling or emptying with air.

■ **symbiosis** Association between two organisms of different species in which both partners benefit. *See also* **commensalism**; **epiphyte**; **parasitism**; **saprophyte**. [2.5.a] [2.6.a]

■ **symmetry** The property of being symmetrical, *i.e.* having the same shapes on each side of or around a point, axis or plane. Most animals have bilateral symmetry; many echinoderms (*e.g.* starfish) have radial symmetry.

■ **sympathetic nervous system** Branch of the **autonomic nervous system** that is structurally different from the **parasympathetic nervous system. Noradrenaline** is produced at the end of sympathetic nerve fibres, unlike the parasympathetic system. Effects produced by each system are generally antagonistic. [3.7.c]

symplast All the living **protoplasm** within a plant, consisting of the individual cells, and the network of filaments that pierce the non-living cell walls and link adjoining cells.

■ **synapse** Point of connection between **neurones** (nerve cells). It consists of a gap between the **membranes** of two cells, across which **impulses** are passed by a transmitter substance (*e.g.* **acetylcholine**). Specialized synapses occur at nerve-**muscle** junctions. *See also* **neurotransmitter**. [3.7.c]

synergy Collective action of two or more things (*e.g.* drugs, muscles) that is more effective than it would be if they acted on their own. Alternative name: synergism.

■ **synovial fluid** Liquid secreted by a **synovial membrane**. [3.3.a] [3.5.e]

■ **synovial membrane** Lining of the capsule that encloses a joint between bones. It secretes synovial fluid, which acts as a lubricant to prevent friction in the joint during movement. [3.3.a] [3.5.e]

syphon *See* **siphon**.

syrinx Sound-producing organ of a bird.

■ **systole** Phase of contraction of the heartbeat during which blood is forced into the **arteries**. *See also* **diastole**. [3.5.b]

T

■ **tadpole** Larval stage of an amphibian (*e.g.* of a frog, toad or newt).

talus The ankle bone.

tannin Yellow substance of vegetable origin (*e.g.* it is found in tree bark, oak galls and tea), used in tanning hides to make leather.

■ **tarsal** Bone that occurs in the feet of tetrapods. Human beings have seven tarsals in each foot, one of them modified to form the heel bone (calcaneum).

■ **tarsus** *1.* Region of the segmented leg of an **insect**. *2.* The ankle region of the hindlimb of a tetrapod vertebrate.

■ **tartaric acid** HOOC.CH(OH)CH(OH).COOH White crystalline hydroxycarboxylic acid that occurs in grapes and other fruits. It is used in dyeing and printing. Its salts, the **tartrates**, are used as buffers (for controlling acidity) and in medicine (*see* **Rochelle salt**). Alternative names: 2,3-dihydroxybutanedioic acid, dihydroxysuccinic acid.

tartrate Ester or **salt** of **tartaric acid**.

■ **taste** Sense that enables animals to detect flavours, which in mammals involves **taste buds**. [3.2.a]

■ **taste bud** Small sense organ containing **chemoreceptors** for the sense of **taste**, located in the mouth (particularly on the upper surface of the **tongue**).

■ **taxis** Orientation of an organism, involving movement, with respect to a **stimulus** from a specific direction. *See also* **chemotaxis**; **geotaxis**; **phototaxis**; **tropism**.

■ **taxonomy** Study of the **classification** of living organisms. [2.5.b]

■ **tear gas** Volatile substance, usually a **halogen**-containing organic compound, that causes irritation of the eyes and is used in crowd control. Alternative name: lachrymator.

■ **tears** Slightly bactericidal watery liquid secreted by the **lacrimal gland** near the eye.

teeth *See* **tooth**.

■ **telophase** Final phase of cell division that occurs in **mitosis** and **meiosis**. During telophase **chromatids**, on reaching the poles of the cell, become densely packed and the cell divides. In animal cells, the **plasma membrane** constricts. In plants, a wall divides the cell in two. [4.6.a]

telotaxis Animal movement in response to a stimulus detected by any form of specialized sense organ.

■ **telson** Lowest segment of the abdomen in some **arthropods** (*e.g.* the tail of a crab), but which is found in insects only at the **embryo** stage.

■ **tendon** Strong **connective tissue** that attaches **muscles** to **bones**. It consists of **collagen** fibres. *See also* **ligament**. [3.3.a] [3.5.e]

■ **tendril** Long thin extension of a stem or leaf, which usually grows in a spiral, developed by many plants as an aid to climbing. [3.4.a] [3.5.e] [3.8.a]

teratogen Any substance, natural or artificial, that causes abnormal development of an **embryo**.

■ **testa** Protective covering of a **seed**. Dry and fibrous, it is formed from the **nucellus** and **integuments**. [3.4.d]

■ **testes** Plural of **testis**. [3.4.c]

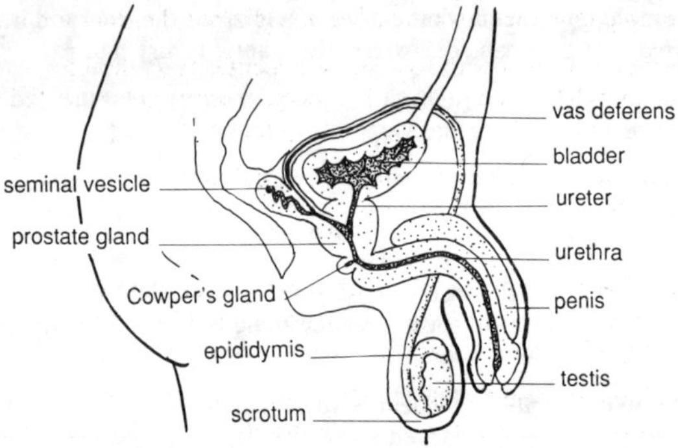

Human testis

■ **testis** Usually paired organ in males of many animals, responsible for the production of **sperm**. In vertebrates, the testes also produce **hormones** that induce development of **secondary sexual characteristics**. [3.4.c]

■ **testosterone Sex hormone** produced in the **testes** that induces the development of **secondary sexual characteristics**. [3.4.c] [3.6.d] [3.9.b]

■ **tetracycline** Member of a class of **antibiotics** (which act by inhibiting **protein** synthesis in bacteria) that have a wide range of applications.

■ **tetrad** *1*. Four cells formed after the second division in

meiosis is complete. *2.* Four **spores** formed by **meiosis** in a spore mother cell, often seen in **fungi**.

■ **tetraploid** Describing an organism that has four sets of homologous **chromosomes**. *See also* **diploid**; **haploid**. [4.7.a]

■ **tetrapod** Animal with four (usually **pentadactyl**) limbs, including all vertebrates with a locomotory system adapted to living on land. Alternative name: quadruped.

■ **thalamus** Part of the **forebrain** of vertebrates that is concerned with the routing of nervous impulses to and from the **spinal cord**.

thalassaemia Inherited form of **anaemia**, common in Mediterranean countries and Africa, caused by a defect in the blood pigment **haemoglobin**.

■ **thalidomide Sedative** drug that is now banned in many countries because it can cause deformities in human embryos. [3.8.b]

■ **Thallophyta** Division of the plant kingdom that contains **algae, bacteria** and **fungi**. The simplest form of plant life, very varied in size and structure, thallophytes all possess a **thallus**. [2.5.b]

■ **thallus** Main body of simple plants (*e.g.* algae, lichens and mosses) that do not have true **roots, stems** or leaves. [3.4.a] [3.5.e] [3.8.a]

thanatosis Defensive behaviour exhibited by some animals (*e.g.* opossums and certain reptiles) that feign death as a means of evading the attention of predators.

thelytoky Sexual adaptation in which **parthenogenesis** is the only form of reproduction in populations that are composed entirely of females, as in *e.g.* the great green cricket.

■ **thermometer, clinical** Mercury thermometer used in medicine for the accurate measurement of body temperature. It measures only a small range of temperatures. A constriction in the capillary near the bulb breaks the mercury thread to retain the temperature reading; before use, the thermometer has to be shaken to return all the mercury to the bulb.

■ **thiamine** White water-soluble crystalline B **vitamin**, found in cereals and yeast. Deficiency of thiamine causes the disorder beri-beri in human beings. Alternative names: thiamin, aneurin, vitamin B_1.

■ **thigmotropism Tropism** in plants (*e.g.* the leaves of *Mimosa pudica* or the tendrils of climbing plants) in response to the **stimulus** of touch.

■ **thorax** *1.* In adult insects, region of the body that bears legs (and wings, if any), situated between head and **abdomen**. *2.* In land-living vertebrates, region of the body that contains the **heart** and **lungs**; the chest. In mammals it is separated from the abdomen by the **diaphragm**. [3.4.a]

threonine Amino acid that is essential in the diet of animals. Alternative name: 2-amino-3-hydroxybutanoic acid.

■ **thrip** Member of the insect order **Thysanoptera**.

■ **thrombin Protein** that takes part in the clotting of **blood** following injury. It converts **fibrinogen** to **fibrin** to help form the clot. [3.5.b]

■ **thrombocyte** Alternative name for a type of blood **platelet**. [3.5.b]

■ **thrombosis** Blockage of a blood vessel by a blood clot (thrombus). If the vessel is an artery serving the heart muscle, thrombosis may result in a heart attack (*i.e.* coronary thrombosis causing a cardiac infarction); if the artery serves the brain, the result may be a **stroke**. [3.5.b]

■ **thrombus** Blood clot. *See* **thrombosis.** [3.5.b]

thymine $C_5H_6N_2O_2$ Colourless crystalline **heterocyclic** compound, one of the **pyrimidines.** It is found in the **nucleotides** of **DNA** (along with **cytosine, adenine** and **guanine**). Alternative names: 5-methyluracil, 5-methyl-2,4-dioxopyrimidine.

thymol $C_{10}H_{14}O$ Colourless crystalline organic compound found in the essential oils of thyme and mint. It is used in antiseptic mouth washes. Alternative names: 2-hydroxy-*p*-cymene, 2-hydroxy-1-isopropyl-4-methylbenzene.

■ **thymus** Twin-lobed **endocrine gland**, situated in the chest near the heart, that plays an important role in the **immune response**. After birth it produces many **lymphocytes** and induces them to develop into **antibody**-producing cells. It declines after puberty and atrophies in older adults. [3.8.b]

■ **thyroid gland** Twin-lobed **endocrine gland** situated in the neck of vertebrates. Under the influence of thyroid-stimulating hormone (TSH) from the **pituitary**, it secretes the **hormone thyroxin**, which is important in growth and **metabolism**. In human beings, undersecretion (hypothyroidism) causes cretinism in children and myxoedema in adults; overproduction (hyperthyroidism) may cause thyrotoxicosis, resulting in goitre. **Iodine** is needed in the diet for efficient functioning of the thyroid; deficiency of iodine may cause the gland to swell and form a goitre. [3.8.b] [3.9.b]

■ **thyroid-stimulating hormone** (TSH) **Hormone** produced by the **pituitary gland** which stimulates the **thyroid gland** into activity.

■ **thyroxin** White crystalline organic compound, an **iodine**-containing **amino acid** derived from **tyrosine**. It is a **hormone**, secreted by the **thyroid gland**.

■ **Thysanoptera** Order of small (less than 2cm) slender insects characterized by short legs and sucking mouthparts, commonly known as thrips. Many are specific pests on human crops (*e.g.* pea thrips, grain thrips).

■ **tibia** *1.* One of the two bones below the knee in a tetrapod vertebrate (the other is the **fibula**); the shinbone. *2.* One of the segments in the leg of an insect. [3.3.a]

■ **tinea** Ringworm, a skin disorder characterized by raised, roughly circular and discoloured patches, that results from an infection by a **fungus**.

■ **tissue** Amalgamation of **cells** (of usually one type) that perform a specific function. In higher organisms tissues may combine to form a highly specialized **organ**. Tissues are found in animals (*e.g.* muscle, connective tissue) and in plants (*e.g.* parenchyma).

■ **tissue culture** Process by which **cells** or **tissues** are maintained outside the body (in vitro) in a suitable medium. The material is kept at a suitable temperature, **pH** and **osmotic pressure**. The composition of the medium depends on the type of tissue cultured. **Cancer** cells do not show normal cell properties when maintained in this way. [3.8.b]

■ **tissue fluid** Alternative name for **lymph**.

■ **tobacco** Plant grown for its leaves, which are dried and smoked (in a pipe or after being made into cigars or cigarettes); for many people, tobacco smoking is a form of **addiction**. Burning tobacco releases various chemicals which the smoker inhales, and some of these have been linked to lung cancer.

tocopherol Vitamin isolated from plants that increases fertility in rats. Deficiency of it causes wasting of muscles in animals. It has been found to have **antioxidant** activity, and it is important in maintaining **membranes**. Alternative name: vitamin E.

■ **tongue** Muscular organ located in the **buccal cavity** (mouth) of some animals, used for manipulating food (and in human beings involved in speech). [3.7.c]

■ **tonsil** One of a pair of **lymphoid tissue** regions located at the back of the mouth which helps to prevent infection by producing **lymphocytes**.

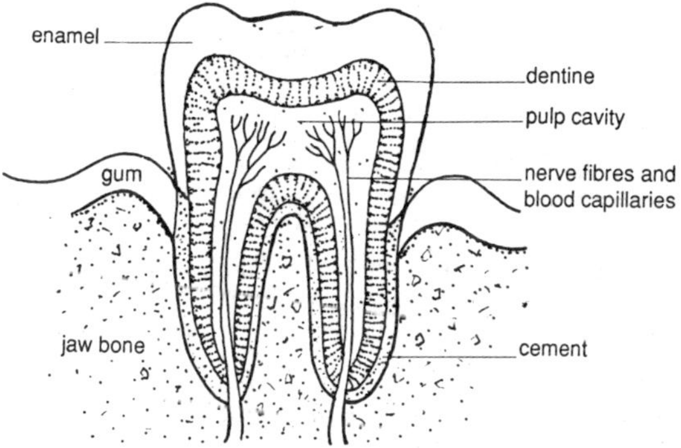

Structure of a molar tooth

■ **tooth** Hard structure embedded in the jawbone and adapted for cutting and grinding food. Teeth are composed of a cavity containing **capillaries** and **nerve** endings, a layer of **dentine** and, over the crown of the tooth, a tough outer layer of **enamel**. *See also* **canine tooth**; **carnassial tooth**; **incisor**; **molar tooth**; **premolar**. [3.4.b] [3.7.b]

■ **toxin** *1.* Poison produced by **bacteria** or other biological sources. *2.* Any harmful substance.

■ **toxoid** Bacterial **toxin** that has been chemically treated to make it non-poisonous, used as a **vaccine** *(e.g.* diphtheria and tetanus).

■ **trace element** Element essential to metabolism, but necessary only in very small quantities (*e.g.* copper and cobalt in animals, molybdenum in plants). Such elements are usually poisonous if large quantities are ingested. [3.7.b] [3.8.b]

■ **trachea** *1.* In mammals, tube through which air is drawn into the **lungs**; the windpipe. *2.* In certain arthropods (*e.g.* insects), one in a system of tubes through which air is drawn into the tissues. [3.4.a] [3.5.e] [3.8.a]

■ **tracheid** Water-carrying elongated plant cell with lignin-containing walls that occurs in **xylem** tissue.

Tracheophyta Major group in the plant kingdom that contains all vascular plants, including the **Pteridophyta** (ferns and their allies) and **Spermatophyta** (seed plants).

■ **trachoma** Disorder that affects the eyes, caused by the bacteria-like micro-organism *Chlamydia trachomatis*. Commonest in tropical Africa and India, it is worldwide the largest single cause of human sight loss.

■ **tranquillizer** Drug that acts on the central nervous system (CNS), used for calming people and animals without affecting consciousness. Some types of tranquillizers, seldom prescribed now, can lead to **addiction**. *See also* **sedative**. [3.4.b] [3.8.b]

transamination Removal and transferance of an **amino group** from one compound (usually an **amino acid**) to another.

transcription Process in cells by which the **genetic code** is copied into single-stranded **messenger RNA** from a **DNA** template. It occurs in the **nucleus** of **eukaryotes** and is separate from **translation** in the **cytoplasm**. In **prokaryotes** these processes are not separate.

■ **transfer RNA** Small molecule of **RNA** that acts as a carrier of specific **amino acids** in the synthesis of **proteins**. Amino acids are placed in a specific order by the transfer RNA molecules according to instructions in the messenger RNA, to form a **polypeptide** chain. [4.9.d]

■ **translation** Process by which **protein** is synthesized in cells. It occurs by the action of **messenger RNA**, which attaches to a **ribosome** in the **cytoplasm. Transfer RNA** molecules which are attached to a specific **amino acid** then line up according to the sequence of amino acids encoded in the messenger RNA to form a **polypeptide** chain. Alternative name: protein synthesis.

■ **translocation** Transport in plants of the soluble products of **photosynthesis**. It occurs in **phloem**. [3.6.b] [3.7.a] [3.9.a]

■ **transpiration** Process by which water is lost from plants by **evaporation**. It occurs mainly through **stomata** in the leaves, and allows movement of water and salts through the plant. It is affected by humidity and temperature, and the drying action of wind. [3.6.b] [3.7.a]

■ **tribe** *1.* In botany, sub-division of a family of plants. *2.* In zoology, local and functional grouping of animals of the same species. *See also* **race; variety**.

■ **tricarboxylic acid cycle** Alternative name for **Krebs cycle**.

■ **triceps** Extensor muscle in the upper arm with three points of origin. It is one of an antagonistic pair of muscles, the other being the biceps.

triglyceride Ester of **glycerol**, in which all three **hydroxyl groups** have been substituted by ester groupings from **fatty acids**. Many **fats** are triglycerides.

triiodomethane CHI_3 Yellow crystalline solid **haloform**, used as an antiseptic. Alternative name: iodoform.

■ **trilobite** Extinct marine animal, a segmented **arthropod** known only from fossils, that externally resembled a large (3–9cm long) woodlouse, and which flourished in great numbers between about 550 million and 350 million years ago.

■ **triploblastic** Describing an animal that has three layers of cells: **ectoderm**, **endoderm** and **mesoderm**. Most members of the **Metazoa** are triploblastic (exceptions include coelenterates, which are diploblastic).

trisaccharide Carbohydrate consisting of three joined **monosaccharides**.

■ **trisomy 21** A state in which a **diploid** cell nucleus has three **chromosomes** of one type. **Down's syndrome** is an example in human beings. [4.6.b] [4.8.b]

■ **trophic level Energy** level found in the **food chain** of an **ecosystem**. Green plants constitute the first trophic level because they are the producers; higher levels take energy from the preceding ones and contain the consumers. The number of organisms at each trophic level becomes smaller as the number of levels increases, **carnivores** being in the highest trophic levels. *See* **pyramid of numbers**. [2.7.b]

■ **tropism** Growth movement exhibited by plants that occurs in response to a specific **stimulus**, *e.g.* light. *See also* **geotropism**; **phototropism**; **taxis**. [3.9.b] [4.9.c]

■ **trypsin** Digestive **enzyme** secreted into the **small intestine**, where it catalyses the **hydrolysis** of **polypeptide** chains (of proteins) at specific sites. [3.7.b]

tryptophan Essential amino acid that contains an aromatic group (*see* **aromatic compound**), needed in animals for proper growth and development.

■ **TSH** Abbreviation of **thyroid-stimulating hormone**.

■ **tube-feet** Arrangement of hollow tubes powered by the water vascular system that open onto the undersides of certain marine invertebrates such as starfish and sea-urchins, and which provide traction during movement. Alternative name: podia. [2.4.a]

■ **tuber** Food-storage organ (*e.g.* in dahlias, potatoes) which develops from a plant stem or root (in which case it is called a root tuber), and which can form buds from which new plants develop (a type of **vegetative propagation**). [3.4.d]

■ **tumour** Abnormal growth, which may be **benign** (*e.g.* a polypus) or **malignant** (*i.e.* cancerous). [3.8.e]

■ **Tunicata** Animals that belong to a subphylum of the **Chordata**; tunicates or sea-squirts. Alternative name: Urochordata.

■ **turgid** Describing something in a state of **turgor**.

■ **turgor** Inflation of plant cells by cell sap (brought about by **osmosis**), which provides the plant with rigidity and internal support. An inflated cell is said to be turgid.

■ **twins** Two offspring born to an animal that normally has only one. In maternal (or identical) twins, two embryos develop when a single fertilized ovum (egg) splits in two. In fraternal (non-identical) twins, two embryos arise when two separate ova are produced and both are fertilized. Maternal twins have the same sex and same genetic make-up; fraternal twins are no more alike than any two offspring of the same parents. [3.4.c]

■ **tympanum** Membrane between the outer **ear** and inner ear. Alternative name: ear drum. [14.6.b]

tyrosine White crystalline organic compound, a naturally occurring **essential amino acid** found in most **proteins**, and a precursor in the body of various **hormones**.

U

■ **ulna** Rearmost (and usually larger) of the two bones in the lower forelimb of a tetrapod vertebrate (the other bone is the radius). [3.5.e]

ultramicroscope Instrument for viewing sub-microscopic objects, *e.g.* bacteria. It is more powerful than an ordinary optical microscope, but not so powerful as an electron microscope.

■ **umbilical cord** *1.* In embryology, vascular structure that contains the **umbilical arteries** and **veins**, connecting the **foetus** to the **placenta**. *2.* In space engineering, flexible and easily disconnectable mechanical or electrical cable which might carry oxygen, information or power to a missile or spacecraft before launching. In space it connects a space-walking astronaut to the spacecraft. [3.4.c]

■ **ungular** Describing a hoof, claw or nail.

■ **ungulate** Hoofed grazing mammal (a **herbivore**), characterized by having hoofs, grinding teeth and often horns or antlers and a comparatively long neck.

■ **unicellular** Describing an organism that consists of only one **cell** (*e.g.* **protozoans**, **bacteria**).

■ **unisexual** Describing an organism with either male or female sex organs, but not both. [2.4.a]

■ **upwelling** Rising to the surface of large lakes and seas of nutrient-rich cold water from lower depths, as a result of currents or wind action. Where this occurs, *e.g.* off the coast of Peru, it sustains a profusion of marine life.

uracil $C_4H_4N_2O_2$ **Pyrimidine base** that forms an essential part of **ribonucleic acid** (RNA). Alternative name: 2,6-dioxypyrimidine.

■ **urea** H_2NCONH_2 White crystalline organic compound, found in the **urine** of mammals as the natural end-product of the metabolism of **proteins**. It is used in plastics, adhesives, fertilizers and animal-feed additives. Alternative name: carbamide. [3.5.e] [3.8.a]

■ **ureter** One of a pair of ducts that carry **urine** from the **kidneys** to the **bladder** of mammals or the **cloaca** of reptiles and birds. It is functionally replaced by the Wolffian duct in fish and amphibians. [3.5.e] [3.8.a]

■ **urethra** Tube through which urine is discharged to the exterior from the urinary bladder of most mammals. [3.5.e] [3.8.a]

■ **uric acid** $C_5H_4N_4O_3$ White crystalline organic acid of the **purine** group, the end-product of the metabolism of **amino acids** in reptiles and birds. In human beings uric acid deposition in the joints is the principal cause of gout. Alternative name: 2,6,8-trihydroxypurine.

■ **urine** Liquid, produced in the **kidneys** and stored in the urinary **bladder**, that contains **urea** and other excretory products. It is discharged to the outside via the **urethra** or **cloaca**. [3.8.a]

Urochordata Subphylum of animals in the phylum **Chordata**; sea squirts. Alternative name: Tunicata.

■ **uterus** Muscular organ located in the lower abdomen of female mammals, in which a fertilized **ovum** develops into a **foetus** prior to birth. Alternative name: womb. [3.4.c]

■ **uvula** Finger-like projection that hangs from the soft palate at the back of the mouth.

V

■ **vaccination** Process of giving somebody a **vaccine**. [3.8.b]

■ **vaccine** Suspension of killed or weakened **antigens** (such as **viruses** or **bacteria**), which is used for immunization. It is either injected (by inoculation) or ingested into the body where it stimulates the production of **antibodies** and so confers **immunity** against infection; both methods are examples of vaccination. Less commonly, vaccines are used in treating a disease. [3.8.b]

vacuity Any gap between the bones of a skull.

■ **vacuole** Alternative name for a **vesicle**. *See also* **contractile vacuole**.

vadose zone Levels of soil above the water table that are aereated, and within which most soil chemistry takes place.

vagility Ability of an organism to adapt to changed environmental circumstances, rated on a scale between 'high' and 'low'.

■ **vagina** *1*. In most female mammals, muscular duct that extends from the **uterus** to the vulvar opening. It receives the male **penis** during mating. *2*. In plants, expanded sheath-like structure at a leaf base. [3.4.c]

■ **vagus nerve** In vertebrates, tenth **cranial nerve**, which forms the major nerve of the **parasympathetic nervous system**, supplying motor nerve fibres to the stomach, kidneys, heart, liver, lungs and other organs. [3.7.c]

valine $C_5H_{11}NO_2$ One of the **essential amino acids** required for normal growth in animals. Alternative names: 2-aminoisovaleric acid, 2-amino-3-methylbutyric acid.

■ **valve** *1.* In botany, part of a dehiscing fruit wall or **capsule**. *2.* In anatomy, a flap of tissue in the body that controls movement of fluid through a tube, duct or aperture in one direction, *e.g.* as between the chambers of the **heart** or in the **veins**.

■ **variation** In biology, differences between members of the same species, which may be either continuous (having a normal distribution about a species mean, *e.g.* height and weight) or discontinuous (having different specific characteristics with no intermediate forms, *e.g.* blood types). *See also* **genetic variation**. [4.4.a] 4.6.a]

■ **variety** Any sub-division of a **species**, *e.g.* breed, race, strain, etc. [2.4.a]

variola Any of the diseases caused by the poxviruses. Variola major is an alternative name for smallpox.

■ **vascular bundle** Strand of fluid-conducting plant tissue that consists of cells which comprise the **xylem** and **phloem**. Separate vascular bundles are linked by interfascular cambium. [3.4.a] [3.5.e] 3.7.a] [3.8.a]

vascular plant *See* **Tracheophyta**.

■ **vascular system** *1.* In animals, system of interlinked fluid-filled vessels, *e.g.* the **blood vascular system**. *2.* In seed plants and **pteridophytes**, system of conducting tissue consisting of **vascular bundles** that is responsible for the transport of mineral salts and water from the roots to the aerial parts of plants, and of food from the leaves to the growing points or to storage organs. It also provides mechanical support in plants with **secondary growth**.

■ **vascular tissue** Composite plant tissue, found in **angiosperms**, **gymnosperms** and **pteridophytes**, that consists of **xylem** and

phloem. It also contains **parenchyma** and **sclerenchyma fibres**, which together form the support tissues. All vascular plants have primary vascular tissue which is formed from the **procambium**. Only vascular plants with **secondary growth** have secondary vascular tissue, formed from the **cambium**.

■ **vas deferens** One of a pair of ducts that carry sperm from the **testes**. In mammals it joins the **urethra** and passes along the **penis**. Plural: vasa deferentia. Alternative name: sperm duct. [3.4.a] [3.4.c] [3.5.e]

■ **vasectomy** Method of sterilizing a male by surgically cutting (and tying the cut ends) of each **vas deferens**, thus making it impossible for sperm from the testes to reach the urethra and penis. *See also* **salpingectomy**. [3.4.c]

■ **vasoconstriction** Reduction in diameter of a **blood vessel** due to contraction of the **smooth muscles** in its walls. It may be induced by the secretion of **adrenaline** in response to pain, fear, decreased blood pressure, low external temperature, etc. or result from stimulation by vasoconstrictor nerve fibres.

■ **vasodilation** Increase in diameter of small **blood vessels** due to relaxation of the **smooth muscles** in their walls. It is induced in response to exercise, high blood pressure, high external temperature, etc. or results from stimulation by vasodilator nerve fibres. Alternative name: vasodilatation.

■ **vasomotor nerve** Nerve of the **autonomic nervous system** that controls the variation in the diameter of **blood vessels**, *e.g.* causing them to become constricted or dilated.

■ **vasopressin Peptide hormone**, secreted by the **pituitary gland** and **hypothalamus**, that stimulates water resorption in the **kidney** tubules, contraction of the **smooth muscles** in the walls of **blood vessels** and permeability of skin and bladder cells in **amphibians**. Alternative name: antidiuretic hormone (ADH).

vauqueline Alternative name for **strychnine**.

■ **vector** In medicine, agent capable of mechanical or biological transference of **pathogens** from one organism to another; *e.g.* the *Anopheles* mosquito is the vector of the malaria parasite. Alternative name: carrier. [2.5.d] [2.6.a]

■ **vegetative propagation** *1.* Method of **asexual reproduction** in plants in which a new plant grows from a part of the 'parent' plant – *e.g.* from a bulb, corm, rhizome, stolon (runner) or tuber. Alternative name: vegetative reproduction. *2.* Asexual reproduction in animals – *e.g.* **budding** in coelenterates such as hydra. Alternative name: vegetative reproduction. [3.4.d]

■ **vein** *1.* In plants, any of several **vascular bundles** in a leaf. *2.* In insects, any of several chitinous tubes that provide membranous wings with support and shape. *3.* In animals, blood vessel that, with the exception of the **pulmonary vein,** carries deoxygenated blood away from cells and tissues. [3.4.a] [3.5.b] [3.5.e] [3.8.a]

velamen Layer of translucent water-retaining spongy cells that surround the **aerial roots** of **epiphytes**.

■ **vena cava** Collective term for the precaval (anterior vena cava) and postcaval (posterior vena cava) vein. The precaval vein is paired, and carries deoxygenated blood away from the head and forelimbs (or arms); the postcaval vein is single and carries deoxygenated blood away from most of the body and hind limbs (or legs) to the heart. [3.4.a] [3.5.b] [3.5.e] [3.8.a]

■ **venation** Arrangement of veins in a plant's leaves or an insect's wings.

■ **ventral** Describing something that is on or near the undersurface of an organism and directed downward (on a human being it is directed forward). *See also* **dorsal**.

■ **ventricle** *1.* In mammals, thick-walled muscular lower chamber of the **heart**. Contraction of the right ventricle pumps deoxygenated blood into the **pulmonary artery**, and contraction of the left ventricle forces oxygenated blood into the **aorta**. *2.* In vertebrates, one of the fluid-filled interconnected cavities within the **brain**. [3.4.a] [3.5.b] [3.5.e] [3.8.a]

■ **venule** *1.* In animals, small **vein** located close to **capillary blood vessels**, where it collects and conveys deoxygenated blood from the capillary network to a **vein**. *2.* In plants, small vein of a leaf. [3.5.b]

■ **vermiform appendix** Alternative term for the **appendix**.

vernation Arrangement of the outer leaves that enclose a developing bud.

■ **vertebra** In **vertebrates**, one of the bones that form the **vertebral column**. Each vertebra typically consists of a solid block of bone (centrum) and a neural arch, which protects the **spinal cord**. In humans there are dorsal and lateral processes (projections) for the attachment of muscles. The spine is divided (from top to bottom) into cervical, thoracic and lumbar vertebrae.

■ **vertebral column** Flexible column of closely arranged **vertebrae** that form an axial **skeleton** running from the **skull** to the tail. It provides a protective channel for the **spinal cord**. The vertebral column becomes larger and stronger towards the posterior (in humans the lower end), which is the major weight-bearing region. Alternative names: spinal column, backbone. [3.3.a] [3.5.e]

■ **Vertebrata** Major subphylum of **Chordata** that contains all animals with a **vertebral column**, *i.e.* mammals, birds, fish, reptiles and amphibians. Vertebrates are characterized by a well-developed **brain**, complex **nervous system** and a flexible **endoskeleton** of **bone** and **cartilage**. Alternative name: Craniata. [2.3.b] [2.4.a] [2.5.b]

■ **vertebrate** Backboned animal; a member of the subphylum **Vertebrata**. [2.3.b] [2.4.a] [2.5.b]

vesicant Blister-causing agent (*e.g.* mustard gas).

■ **vesicle** In biology, small fluid-filled sac of variable origin, *e.g.* **Golgi apparatus**, pinocytotic vesicle. Alternative names: **vacuole**, air sac, **bladder**.

■ **vessel** *1.* In animal anatomy, tubular structure that transports fluid (*e.g.* blood, lymph). *2.* In seed plants, advanced form of **xylem** made up of vessel elements.

■ **vestigial organ** Reduced structure that has lost its original function during the course of **evolution**, but resembling the corresponding fully functional organs found in a related organism, and hence manifesting an evolutionary relationship, *e.g.* wings of flightless birds, limb girdles of snakes. [4.9.a]

■ **Vibrio** Type of Gram-positive comma-shaped **bacterium** which includes the organism that causes cholera.

vicariance Evolution of two closely related species from populations of the same species that became separated geographically. There are many examples between the Old World (Europe and Africa) and the New World (the Americas).

■ **villus** One of many finger-like structures that line the inside of the **small intestine**. Villi increase the surface area for absorption. Each villus contains a central **lacteal** and a network of blood **capillaries**, which absorb the soluble products of **digestion** into the body. [3.5.c] [3.7.b]

viroid Smallest disease-causing agent, a tight loop of **RNA** lacking any form of capsid (outer coat).

■ **virus** Pathogenic micro-organism with a diameter between about 20 and 400 nm, visible only under an electron microscope. It consists of an outer coat (capsid) of **protein** and an inner core of deoxyribonucleic acid **(DNA)** or ribonucleic acid **(RNA)**. Viruses infect plants, animals and bacteria. Outside the host cells viruses are metabolically inactive (and may be crystallizable), and only when attached to a cell or wholly inside it does the viral DNA interfere with the metabolic activities of the cell, suppressing its normal control processes and causing it to manufacture new protein coats and nucleic acid threads identical with those of the invading virus. Viruses which actively attack and proliferate in cells are described as virulent. [3.8.b] [3.8.c]

■ **viscera** Internal organs, particularly the gut (intestines). [3.4.a]

■ **visceral** Relating to the **viscera**.

■ **visual purple** Alternative name for **rhodopsin**. [15.6.b]

■ **vitamin** Any of several organic compounds that in small quantities are essential to the proper growth and regulation of metabolic processes, *e.g.* energy transformation in animal organisms. There are two major groups, water-soluble (*e.g.* vitamins C, B) and fat-soluble (*e.g.* A, D, E, K), which are present in foodstuffs and must be taken as part of a balanced diet. [3.7.b]

■ **vitamin A** *See* **retinol**.

■ **vitamin B_1** *See* **thiamine**.

■ **vitamin B_2** *See* **riboflavin**.

■ **vitamin C** *See* **ascorbic acid**.

■ **vitamin D** *See* **calciferol**.

vitamin E *See* **tocopherol**.

vitamin H *See* **biotin**.

■ **vitreous humour** In the vertebrate **eye**, firm, transparent gel-like substance that fills the the space behind the **lens**, thus maintaining the shape of the eyeball. *See also* **aqueous humour**. [15.6.b]

■ **viviparous** *1.* In animals, giving birth to live young rather than laying eggs which hatch later. This ability is almost entirely confined to mammals, although the term is generally used of animals other than mammals in which viviparity is unusual, *e.g.* certain snakes. *2.* In plants, having seeds that germinate and start growing within the fruit, *e.g.* mangrove. [3.8.a]

■ **vocal cord** One of a pair of membranous flaps in the **larynx** that are vibrated by air from the lungs to produce sounds.

■ **voluntary muscle** Type of muscle, connected to the bones, that is under conscious control. It is responsible for most body movements. Alternative name: striated muscle. [3.5.e]

■ **volvox** One of a group of ciliated freshwater **algae** (genus *Volvox*) that form well-organized spherical colonies consisting of thousands of cells. [2.4.a]

■ **vulva** In female mammals, the external opening of the **vagina**. [3.4.a] [3.4.c] [3.5.e] [3.8.a]

W

■ **Wallace's line** Imaginary line between Australia/New Guinea and south-eastern Asia that separates the native mammals of Australasia (monotremes and marsupials) from those of Asia (placentals). It was named after the British naturalist Alfred Russel Wallace (1823–1913). [4.9.a]

■ **warfarin** Organic compound used (as its sodium derivative) as an anticoagulant drug and as a pesticide for killing rats and mice.

■ **warm-blooded** Alternative name for **homoiothermic**. [2.4.a]

■ **warning coloration** Bright patterns that occur in some noxious animals (especially insects) which results in a predator learning to avoid them. Alternative name: aposematic coloration. [2.5.d]

■ **water cycle** Continuous movement of water between the atmosphere and the land and oceans. Water that falls as precipitation (*e.g.* rain, snow, hail) passes into the ground or runs off to form springs and rivers, which ultimately flow into the oceans. Water evaporates from the oceans as vapour, forms clouds in the atmosphere, and falls again as precipitation. Some is also returned to the atmosphere through transpiration and respiration after having been taken in by plants and animals. Alternative name: hydrological cycle. [5.6.a] [9.5.c]

■ **wax** Solid or semi-solid organic substance that is *1.* an **ester** of a **fatty acid**, produced by a plant or animal (*e.g.* beeswax, tallow), or *2.* a high molecular weight hydrocarbon (*e.g.* paraffin wax, from petroleum), also called mineral wax.

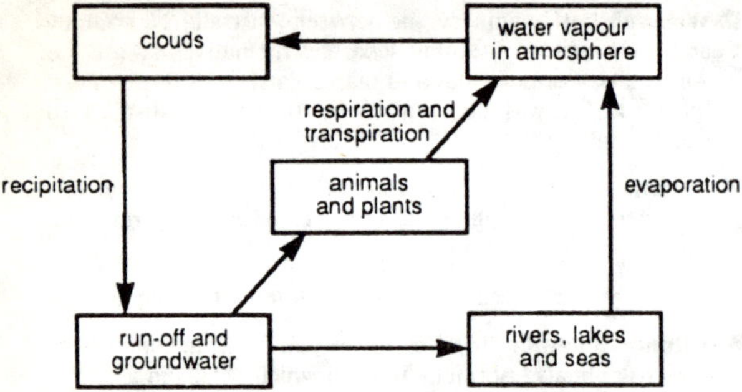

The water cycle

■ **wildlife** Any non-domesticated organisms, although the term is often used to refer only to wild animals.

■ **wilting** Limpness of plant tissue that occurs when there is insufficient cell sap to keep the cells rigid, and which may be caused by lack of water or disease. *See also* **flaccid**. [3.4.a] [3.5.e] [3.7.a] [3.8.a]

wood sugar Alternative name for **xylose**.

■ **wool fat** Alternative name for **lanolin**.

■ **wound-tumour virus** One of a group of **viruses** that cause the formation of galls (*e.g.* oak apples) on many plants, and which are often transmitted by insects.

X

xanthene $CH_2(C_6H_4)_2O$ Yellow crystalline organic compound, used as a fungicide and in making dyes. Alternative name: tricyclicdibenzopyran.

xanthone $CO(C_6H_4)_2O$ Plant **pigment** that occurs in gentian and other flowers, used commercially as an insecticide and dye intermediate. Alternative name: 9H-xanthen-9-one.

■ **xanthophyll** $C_{40}H_{56}O_2$ Yellow to orange **pigment** present in the normal **chlorophyll** mixture of green plants. [3.6.b] [3.7.a] [3.9.a]

■ **X-chromosome Sex chromosome** that occurs paired in human females (*i.e.* as XX), and coupled with a **Y-chromosome** (XY) in males. It is the larger of the two sex chromosomes, and carries many **sex-linked genes**. [4.7.a]

xeric Describing environments that are predominantly dry.

■ **xerophyte** Plant that grows in dry habitats. Adaptations that enable it to do this include swollen stems that store water and leaf modifications (copious hairs, a thick cuticle, reduction of leaves to spines as in cacti) that reduce loss of water. [2.6.a]

xerosere Pattern of plant **succession** that is characteristic of arid environments.

■ **X-rays** Electromagnetic (and ionizing) radiation produced in partial vacuum by the sudden arrest of high-energy bombarding electrons as they hit the heavy atomic nuclei of a target metal. They have very short wavelengths and can penetrate solids to varying degrees. This characteristic has made them useful in carefully controlled doses in medicine, dentistry and X-ray crystallography. Overexposure to X-rays can be harmful, possibly causing cancer.

■ **xylem** In higher plants, water-conducting tissue that consists of lignified vessel elements, **tracheids**, fibre and **parenchyma** tissue, which together provide mechanical support. It is the woody part of a plant. [3.4.a] [3.5.e] [3.8.a]

xylose $C_5H_{10}O_5$ Naturally occurring **pentose** sugar, found in the form of xylan or as **glycosides** in many plants (*e.g.* cherry and maple wood, straw, pecan shell, corn cobs and cottonseed hulls). Alternative name: wood sugar.

Y

yaws Tropical ulcerative disease caused by the bacterium *Treponema pertenue*, characterized by raspberry-like skin eruptions. Alternative name: framboesia.

■ **Y-chromosome** Smaller of the two **sex chromosomes**, found only in the **heterogametic** sex; *e.g.* human males have one Y-chromosome and one X-chromosome (XY). [4.7.a]

■ **yeast** Collective name for **fungi** whose vegetative bodies consist of single individual cells. Yeasts contain **enzymes** (*e.g.* **zymase**) which bring about **fermentation**. They are used in making bread (when they ferment sugar to produce carbon dioxide gas, which causes the bread to rise) and in making wine and beer (when they ferment sugar to produce alcohol). [7.5.b]

■ **yellow spot** (macula lutea) Alternative name for **fovea**. [15.6.b]

■ **yolk** Part of an ovum that stores the nutritive materials, or the yellow central portion of the egg of birds and reptiles. [3.4.c]

Z

zooid Individual invertebrate animal that lives as part of a **colony**.

■ **zoology** Systematic study of animals with relation to other animals, plants and their non-living environment.

zooneuston Narrow layer above and below the surface of open water containing organisms that exploit the effects of **surface tension**, *e.g.* pond skaters.

zoonosis Disease that can be transmitted from animals to human beings (*e.g.* rabies).

■ **zooplankton** In an aquatic **ecosystem**, group of passively floating and drifting microscopic animals. *See also* **plankton**.

■ **zygote** **Diploid cell** that results from the fertilization of a female **gamete** by a male gamete. It sometimes undergoes immediate cleavage, but may also develop a thick resistant outer coat to form a zygospore. [4.9.d]

zymase **Enzyme** that catalyses the **fermentation** of **carbohydrates** to ethanol (ethyl alcohol).

zymogen Inactive precursor of an **enzyme** formed by plants and animals. It is activated by the action of a **kinase** or **zymoexciter**. Alternative name: proenzyme.

zymotic Describing an agent that causes an infectious disease.

APPENDIX I

Animal Classification (major phyla)

Phylum	*Subphylum*	*Class*
Protozoa	Ciliophora	Kinetofragminophora Oligohymenophora Polyhymenophora
	Cnidospora	Microsporidea Myosporidea
	Sarcomastigophora	Actinopodea Phytomastigophora Rhizopodea Zoomastigophora
	Sporozoa	Piroplasmea Sporozea
Porifera		Calcaria Demospongiae Hexactinellida Sclerospongiae
Cnidaria		Anthozoa Hydrozoa Scyphozoa
Ctenophora		Nuda Tentaculata
Platyhelminthes		Cestoda Trematoda Turbellaria

Phylum	*Subphylum*	*Class*
Rhyncocoela		Anopla Enopla
Rotifera		Digononta Monogonta
Nematoda		Aphasmida Phasmida
Nematomorpha		Gordioidea Nectonematoidea
Annelida		Hirudinea Oligochaeta Polychaeta
Echinodermata		Asteroidea Crinoidea Echinoidea Holothuroidea Ophiuroidea
Mollusca		Alplacophora Bivalvia Cephalopoda Gastropoda Monoplacophora Polyplacophora Scaphopoda
Arthropoda	Chelicerata	Arachnida Merostomata Pycnogonida
	Crustacea	Branchiura Branchipoda Cephalocarida

Phylum	*Subphylum*	*Class*
	Crustacea (cont.)	Cirripedia Copepoda Mystacocarida Ostracoda
	Uniramia	Chilopoda Diplopoda Insecta Pauropoda
Chordata	Cephalochordata	
	Hemichordata	Enteropneusta Pterobranchia
	Vertebrata	Agnatha Elasmobranchii Osteichthyes Amphibia Reptilia Aves Mammalia
	Urochordata	

APPENDIX II

Plant Classification (major divisions)

Division	*Subdivision*	*Class*
(Prokaryotic algae)		
Cyanophyta		
Prochlorophyta		
(Eukaryotic algae)		
Bacillariophyta		
Chlorophyta		
Chrysophyta		
Cryptophyta		
Euglenophyta		
Phaeophyta		
Pyrrophyta		
Rhodophyta		
Xanthophyta		
Mycota	Eumycotina	Ascomycetes
		Basidiomycetes
		Chytridiomycetes
		Deuteromycetes
		Hypochytridiomycetes
		Oomycetes
		Plasmodiophoromycetes
		Trichomycetes
		Zygomycetes
	Myxomycotina	Myxomycetes

Division	*Subdivision*	*Class*
Lichenes		Ascolichenes Basidiolichenes
Bryophyta		Anthoceratopsida Bryopsida Hepaticopsida
Pteridophyta		Filicopsida Lycopsida Psilopsida Sphenopsida
Spermatophyta	Gymnospermae	Coniferopsida Cycadopsida Gnetopsida
	Magnoliophyta	Liliopsida Magnoliopsida

APPENDIX III

Nobel Prizewinners in Physiology or Medicine

Some achievements in biology are recognized by the award of a Nobel Prize in Chemistry (whose award-winners are listed in an appendix to the companion *Chemistry Mini Dictionary*). Most award-winning biologists, however, receive the prize for Physiology or Medicine.

1901	E. von Behring (German): development of diphtheria antitoxin
1902	R. Ross (British): malaria and its transmission
1903	N. Finsen (Danish): light in the treatment of disorders
1904	I. Pavlov (Russian): study of digestion
1905	R. Koch (German): tuberculosis and its transmission
1906	C. Golgi (Italian) and S. Ramón y Cajal (Spanish): study of nerves
1907	C. Laveran (French): protozoal disorders
1908	P. Ehrlich (German) and E. Metchnikoff (Russian/French): study of immunity
1909	E. Kocher (Swiss): thyroid gland and its function
1910	A. Kossel (German): chemistry of cells
1911	A. Gullstrand (Swedish): light diffraction by the eye
1912	A. Carrel (French): methods for grafting vessels and organs
1913	C. Richet (French): study of allergies
1914	R. Bárány (Austrian): the balancing mechanism of the inner ear
1915–18	*No award*
1919	J. Bordet (Belgian): study of immunity

1920	A. Krogh (Danish): discovering blood capillary action
1921	*No award*
1922	A. Hill (British) and O. Meyerhof (German): discovery of heat production and lactic acid formation in muscles
1923	F. Banting (Canadian) and J. Macleod (British): discovery of insulin
1924	W. Einthoven (Dutch): invention of electrocardiograph
1925	*No award*
1926	J. Fibiger (Danish): induction of cancer using parasites
1927	J. Wagner-Jauregg (Austrian): treatment of paralysis using hyperthermia
1928	C. Nicolle (French): study of typhus
1929	C. Eijkman (Dutch) and F. Hopkins (British): discovery of vitamins connected with beriberi and growth
1930	K. Landsteiner (American): discovery of ABO blood types
1931	O. Warburg (German): role of enzymes in tissue respiration
1932	E. Adrian and C. Sherrington (British): study of neurone function
1933	T. Morgan (American): role of chromosomes in heredity
1934	G. Minot, W. Murphy and G. Whipple (American): use of liver in the treament of anaemia
1935	H. Spemann (German): discovery of organizing centres for embryonic development
1936	H. Dale (British) and O. Loewi (German/American): discovery of chemical neurotransmitters

1937	A. Szent-Györgyi (Hungarian): study of tissue oxidation, vitamin C, etc.
1938	C. Heymans (Belgian): study of regulation of respiration
1939	G. Domagk (German): sulphonamide drugs (award presented in 1947)
1940–42	*No award*
1943	H. Dam (Danish) and E. Doisy (American): discovery and synthesis of vitamin K
1944	J. Erlanger and H. Gasser (American): study of single nerve fibres
1945	A. Fleming, H. Florey and E. Chain (British): discovery and drug use of penicillin
1946	H. Muller (American): mutagenic effect of X-rays
1947	C. and G. Cori (American) and B. Houssay (Argentinian): studies of insulin, the pancreas and the pituitary
1948	P. Müller (Swiss): discovery of DDT insecticide
1949	W. Hess (Swiss) and A. Moniz (Portuguese): study of various brain areas and their functions
1950	P. Hench, E. Kendall (American) and T. Reichstein (Swiss): function of cortisone and ACTH
1951	M. Theiler (S. African/American): development of yellow fever vaccine
1952	S. Waksman (American): discovery of streptomycin
1953	F. Lipmann (German/American) and H. Krebs (German/British): discoveries in biosynthesis (Krebs cycle)
1954	J. Enders, T. Weller and F. Robbins (American): in vitro method of producing polio virus
1955	H. Theorell (Swedish): studies of oxidases (oxidation enzymes)
1956	A. Cournand, D. Richards Jr (American) and W. Forssmann (W. German): cardiac catheterization

1957 D. Bovet (Italian): discovery of antihistamines
1958 G. Beadle, E. Tatum and J. Lederberg (American): biochemical and bacterial genetics
1959 S. Ochoa and A. Kornberg (American): artificial synthesis of nucleic acid
1960 M. Burnet (Australian) and P. Medawar (British): research on organ transplants
1961 G. von Békésy (Hungarian/American): mechanism of sound discrimination by the ear
1962 F. Crick and M. Wilkins (British) and J. Watson (American): composition of nucleic acids
1963 J. Eccles (Australian), A. Hodgkin and A. Huxley (British): transmission and behaviour of nerve impulses
1964 K. Bloch (German/American) and F. Lynen (W. German): study of cholesterol and fat metabolism
1965 F. Jacob, A. Lwoff and J. Monod (French): genetic control of synthesis of enzymes
1966 C. Huggins and F. Rous (American): discovery of virus-induced cancer and hormone treatment for cancer
1967 R. Granit (Swedish), H. Hartline and G. Wald (American): biochemistry and physiology of the eye
1968 R. Holley, H. Khorana and M. Nirenberg (American): explanation of genetic control of cell function
1969 M. Delbruck, A. Hershey and S. Luria (American): research on bacteriophages
1970 J. Axelrod (American), B. Katz (British) and U. von Euler (Swedish): role of chemical nerve transmitters
1971 E. Sutherland Jr (American): hormone action and discovery of cyclic AMP
1972 G. Edelman (American) and R. Porter (British): chemical structure of antibodies

1973	N. Tinbergen (Dutch), K. Lorenz and K. von Frisch (Austrian): studies of animal behaviour
1974	A. Claude, C. de Duve (Belgian) and G. Palade (Romanian/American): studies of cell biology
1975	D. Baltimore, H. Temin (American) and R. Dulbecco (Italian/American): effects of viruses on cancer cell genes
1976	B. Blumberg and D. Gajdusek (American): origin and etiology of slow virus infections
1977	R. Guillemin, A. Schally and R. Yalow (American): role of hormones in body chemistry
1978	W. Arber (Swiss), D. Nathans and H. Smith (American): discoveries in molecular genetics
1979	A. Cormack (American) and G. Hounsfield (British): development of the CAT scanner
1980	B. Benacerraf (Venezuelan/American), G. Snell (American) and J. Dausset (French): genetic control of the immune system
1981	D. Hubel, R. Sperry (American) and T. Wiesel (Swedish): study of brain function and organization
1982	S. Bergstrom, B. Samuelsson (Swedish) and J. Vane (British): study of prostaglandins
1983	B. McClintock (American): anomalous intracellular gene behaviour
1984	N. Jerne (British/Danish), G. Köhler (W. German) and C. Milstein (Argentinian/British): discoveries in immunology
1985	M. Brown and J. Goldstein (American): relationship between excessive cholesterol and heart disease
1986	S. Cohen (American) and R. Levi-Montalcini (Italian/American): studies of growth in cells and organs
1987	S. Tonegawa (Japanese): discoveries on how genes alter to form antibodies against specific antigens

1988 J. Black (British), G. Elion and G. Hitchings (American): discovery of beta-blockers, H-2 blockers and anti-cancer drugs

1989 I. Bishop and H. Varmus (American): discovery of the process that causes abnormal cellular genes to become cancerous

1990 J.E. Murray and E.D. Thomas (American): discoveries relating to organ transplants

APPENDIX IV

The National Curriculum Attainment Targets given in square brackets at the ends of many articles in the Dictionary are those as defined for the current (1991) academic year. However, the Curriculum is under constant revision, and it is intended to change and simplify many of the designations for subsequent years. When these changes are implemented, the following table allows the reader to convert from the current to the future Attainment Targets. A coding not listed in the 'Original' column has no equivalent in the 'New' system.

Original	*New*	*Original*	*New*	*Original*	*New*	*Original*	*New*
2.1.a	2.1.b	3.1.a	2.1.a	4.1.a	2.1.b	5.2.b	2.2.d
2.2.a	2.2.b	3.2.b	2.1.c	4.3.a	2.4.c	5.4.a	2.2.d
2.3.a	2.2.b	3.3.a	2.3.a	4.5.a	2.5.b	5.5.a	2.3.c
2.3.b	2.3.b	3.3.b	2.3.a	4.6.a	2.6.b	5.6.c	2.4.c
2.4.b	2.2.d	3.4.a	2.4.a	4.7.a	2.8.b	5.7.b	2.8.c
2.5.a	2.2.c	3.5.b	2.4.c	4.8.a	2.7.b	5.8.a	2.9.d
2.5.b	2.4.b	3.5.e	2.5.a	4.8.b	2.9.c	5.9.a	2.9.d
2.5.c	2.5.c	3.6.b	2.3.d	4.8.c	2.9.c	5.10.a	2.10.d
2.5.d	2.4.d	3.8.a	2.7.a	4.9.a	2.9.c		
2.6.a	2.6.a	3.10.a	2.9.a	4.9.b	2.9.b		
2.6.b	2.6.d			4.9.c	2.10.a		
2.7.a	2.8.d			4.9.d	2.10.b		
2.7.b	2.7.d			4.10.b	2.10.b		
2.9.a	2.7.c			4.10.d	2.10.c		
2.9.c	2.10.e						
2.10.a	2.10.e						
2.10.b	2.9.e						

HARRAP'S MINI SERIES

A comprehensive, pocket-sized reference library available at your local bookshop.

ENGLISH SERIES

Mini English Dictionary
ISBN 0 245-60140-6

English Grammar
ISBN 0 245-54746-0

English Idioms
ISBN 0 245-60041-8

English Punctuation and Hyphenation
ISBN 0 245-60020-5

English Spelling
ISBN 0 245-54832-7

English Spelling Rules
ISBN 0 245-60144-9

English Synonyms
ISBN 0 245-54831-9

English Usage
ISBN 0 245-54830-0

English Verbs
ISBN 0 245-54745-2

English Vocabulary
ISBN 0 245-54747-9

FOREIGN LANGUAGES

Mini French Dictionary
ISBN 0 245-54507-7

French Grammar
ISBN 0 245-54582-4

French Verbs
ISBN 0 245-54581-6

French Vocabulary
ISBN 0 245-54583-2

Mini German Dictionary
ISBN 0 245-54573-5

German Grammar
ISBN 0 245-54631-6

German Verbs
ISBN 0 245-54629-4

German Vocabulary
ISBN 0 245-54630-8

Mini Spanish Dictionary
ISBN 0 245-54586-7

Spanish Grammar
ISBN 0 245-54690-1

Spanish Idioms
ISBN 0 245-60039-6

Spanish Synonyms
ISBN 0 245-60040-X

Spanish Verbs
ISBN 0 245-54688-X

Spanish Vocabulary
ISBN 0 245-54689-8

Mini Italian Dictionary
ISBN 0 245-54585-9

Italian Grammar
ISBN 0 245-60052-3

Italian Verbs
ISBN 0 245-60053-1

Italian Vocabulary
ISBN 0 245-60051-5